Florian Ion PETRESCU &
Relly Victoria PETRESCU

NEWS IN PHYSICS

Germany 2012

Scientific reviewer:

Dr. Veturia CHIROIU

Honorific member of
Technical Sciences Academy of Romania (ASTR)
PhD supervisor in Mechanical Engineering

Copyright

Title book: News in Physics

Authors book: Florian Ion Petrescu & Relly Victoria Petrescu

© 2001-2012, Florian Ion PETRESCU

petrescuflorian@yahoo.com

ALL RIGHTS RESERVED. This book contains material protected under International and Federal Copyright Laws and Treaties. Any unauthorized reprint or use of this material is prohibited. No part of this book may be reproduced or transmitted in any form or by any means, electronic or mechanical, including photocopying, recording, or by any information storage and retrieval system without express written permission from the authors / publisher.

Manufactured and published by:
Books on Demand GmbH, Norderstedt
ISBN 978-3-8482-2964-2

Welcome! A Short Book Description

The movement of an electron around the atomic nucleus has today a great importance in many engineering fields. Electronics, aeronautics, micro and nanotechnology, electrical engineering, optics, lasers, nuclear power, computing, equipment and automation, telecommunications, genetic engineering, bioengineering, special processing, modern welding, robotics, energy and electromagnetic wave field is today only a few of the many applications of electronic engineering. This first chapter presents shortly a new and original relation which calculates the radius with that the electron is running around the atomic nucleus.

The second chapter presents, shortly, a new and original relation (20) which calculates the Doppler Effect exactly. This new relation (20) is the exact form and the classical expression (10) is an approximate relation.

Renewable energy is energy which comes from natural resources such as sunlight, wind, rain, tides, and geothermal heat, which are renewable (naturally replenished). The share of renewables in electricity generation is around 18%, with 15% of global electricity coming from hydroelectricity and 3% from new renewables. The third chapter aims to disseminate new methods of obtaining energy. After 1950, began to appear nuclear fission plants. The fission energy was a necessary evil. In this mode it stretched the oil life, avoiding an energy crisis. Even so, the energy obtained from oil represents about 66% of all energy used. At this rate of use of oil, it will be consumed in about 40 years. Today, the production of energy obtained by nuclear fusion is not yet perfect prepared. But time passes quickly. We must rush to implement of the additional sources of energy already known, but and find new energy sources. In these circumstances this chapter comes to proposing possible new energy sources, like energies obtained by the annihilation of a particle with its antiparticle.

Scientific reviewer:

Dr. Veturia CHIROIU
Honorific member of
Technical Sciences Academy of Romania (ASTR)
PhD supervisor in Mechanical Engineering

You are welcome to read the full book!

CONTENT

Welcome! A Short Book Description................... 03
Content.. 04
Presentation... 06
Chapter I – Presenting of an atomic model
and some possible applications in laser field,
H-hour...09
Introduction.. 09
Used notations.. 13
Determining the two different electron
speed values.. 13
Determining the masses and the energy of
the atomic electron in movement.................. 17
The possible laser frequencies..................... 18
The laser frequencies and conclusions........... 20
The relationships, Determining the ray of an
electron moving on an orbit around an atom........ 23
Determining the velocities of an electron
which is running around an atom..................... 24
Determining the mass of the
electron in movement.................................. 25
Determining the energy of
the electron in movement............................. 25
Determining the frequencies of pumping............ 25
Conclusions... 27

Correcting the length of the r radius................ 27
H-hour... 30
Cold nuclear fusion.................................. 35
Aneutronic fusion.................................... 36
Dense plasma focus................................... 39
Chapter II – Some few specifications about the Doppler Effect to the electromagnetic waves........42
Introduction... 42
Development.. 43
Application.. 43
Some few specifications about the Doppler Effect to the electromagnetic waves........................... 56
Chapter III – The future energy...................... 62
Obtaining energy by the annihilation of an electron with a positron, or annihilation of a proton with an antiproton (case studies presentation)................63
Chapter IV - New aircraft............................ 67
4.1. Ion thruster.................................... 67
4.2. The hipep engine................................ 71
4.3. New ionic or beam pulses engines................ 72
4.4. Calculation of the momentum of particle jets. 76
4.5. Calculation of the acceleration of the ship...... 77
Chapter V - Capturing energy concentrated near the source and forwarding directly to Earth in concentrated form........................ 78

PRESENTATION

CHAPTER I - PRESENTING OF AN ATOMIC MODEL AND SOME POSSIBLE APPLICATIONS IN LASER FIELD

The movement of an electron around the atomic nucleus has today a great importance in many engineering fields.

Electronics, aeronautics, micro and nanotechnology, electrical engineering, optics, lasers, nuclear power, computing, equipment and automation, telecommunications, genetic engineering, bioengineering, special processing, modern welding, robotics, energy and electromagnetic wave field is today only a few of the many applications of electronic engineering.

With the help of powerful lasers one can create a dense and highly ionized plasma. We need a highly ionized dense plasma to achieve nuclear fusion (cold or hot).

Since 1989, it talks about achieving nuclear fusion hot and cold. Another two decades have passed and humanity still does not benefit from nuclear fusion energy. What actually happens? Is it an unattainable myth? It was also circulated by the media that has been achieved nuclear fusion heat. Since 1989 there are all sorts of scientists with all kinds of crafted devices, which declare that they can produce nuclear power obtained by cold fusion (using cold plasma). May be that these devices works, but their yield is probably too small, or at an enlarged scale these give not the expected results. This is the real reason why we can't use yet the survival fuel (the deuterium).

Unfortunately today the dominant processes that produce energy are combustion (reaction) chemical combination of carbon with oxygen. Thermal energy released from such reactions is conventionally valued at about 7000 calories per gram.

Only the early 20th century physicists have succeeded in producing, other energy than by traditional methods. Energy release per unit mass was enormous compared with that obtained by conventional procedures. The Kilowatt based on nuclear fission of uranium nuclei has today a significant share in global energy balance. Unfortunately, the nuclear power plants burn the fuel uranium, already considered conventional and on extinct.

The current nuclear power is considered a transition way, to the energy thermonuclear, based on fusion of light nuclei.

The main particularity of synthesis reaction (fusion) is the high prevalence of the used fuel (primary), deuterium. It can be obtained relatively simply from ordinary water.is first chapter presents shortly a new and original relation who calculates the radius with that the electron is running around the atomic nucleus.

CHAPTER II – SOME FEW SPECIFICATIONS ABOUT THE DOPPLER EFFECT TO THE ELECTROMAGNETIC WAVES

This chapter presents, shortly, a new and original relation (20) which calculates the Doppler Effect exactly. This new relation (20) is the exact form and the classical expression (10) is an approximate relation.

The classical approximate relation (10) presented in the form (15) can't foresee the Doppler Effect for the case when the angle $\varphi=90^0$. For this reason it was introduced the relativity effect, where the period T_0 take the form T_0/α. Before to utilize the theory of the relativity it's strongly necessary to test the relations (8), (18) or (20), and the particular form (14) (for the angle $\varphi=90^0$), to testing the Doppler exact effect without the relativity theory.

The Doppler Effect represents the frequency variation of the waves, received by an observer which is drawing (coming), respectively it's removing (going), from a wave spring (source).

If a bright spring is drawing to an observer, the frequency of waves received by the observer is bigger than the emitted frequency of source, such that the respective spectral lines are moving to violet. On the contrary, if the light source is removing from the observer, the spectral lines are moving to red.

One proposes to study the Doppler Effect for the light waves, generally for the electromagnetic waves. The paper proposes for the Doppler Effect the relation (20) which can replace the classical form (10).

CHAPTER III - The Future Energy

Renewable energy is energy which comes from natural resources such as sunlight, wind, rain, tides, and geothermal heat, which are renewable (naturally replenished). In 2008, about 19% of global final energy consumption came from renewables, with 13% coming from traditional biomass, which is mainly used for heating, and 3.2% from hydroelectricity. New renewables (small hydro, modern biomass, wind, solar, geothermal, and biofuels) accounted for another 2.7% and are growing very rapidly.

The share of renewables in electricity generation is around 18%, with 15% of global electricity coming from hydroelectricity and 3% from new renewables. This chapter aims to disseminate new methods of obtaining energy. After 1950, began to appear nuclear fission plants.

The fission energy was a necessary evil. In this mode it stretched the oil life, avoiding an energy crisis. Even so, the energy obtained from oil represents about 66% of all energy used. At this rate of use of oil, it will be consumed in about 40 years. Today, the production of energy obtained by nuclear fusion is not yet perfect prepared.

But time passes quickly. We must rush to implement of the additional sources of energy already known, but and find new energy sources. In these circumstances this chapter comes to proposing possible new energy sources, like energies obtained by the annihilation of a particle with its antiparticle.

CHAPTER IV - NEW AIRCRAFT

Speaking about a new ionic engine means to speak about a new aircraft. This chapter presents shortly the actual ionic engines (called ion thrusters) and the new ionic (pulse) engines proposed by the author.

Ionic engine (ion thruster, which accelerates the positive ions through a potential difference) is about 10 times more effective than classic system based on combustion.

We can still improve the efficiency of 10-50 times if one uses pulses of positive ions accelerated in a cyclotron mounted on the ship; the efficiency can easily grow for 1000 times if the positive ions will be accelerated in a high energy synchrotron, synchrocyclotron or isochronous cyclotron (1-100 GeV). In this, the big classic synchrotron is reduced to a ring surface (magnetic core). Future (ionic) engine will have mandatory a circular particle accelerator (high or very high energy).

We can thus increase the speed and autonomy of the ship using a less quantity of fuel and power.

One can use synchrotron radiation (synchrotron light, high intensity beams), like high intensity (X-ray or Gamma ray) radiation, as well. In this case will be a beam engine (not an ionic engine), it'll use only the power (energy, which can be solar energy, nuclear energy, or both) and so we will remove the fuel. It proposes using a powerful LINAC at the exit of synchrotron (especially when one accelerates electrons) to not lose energy by photons premature emission.

With a new ionic engine one builds a new aircraft, which can travel through water and. This new aircraft will can accelerate directly, without an additional combustion engine and without gravity assists from other planets.

CHAPTER V - CAPTURING ENERGY CONCENTRATED NEAR THE SOURCE AND FORWARDING DIRECTLY TO EARTH IN CONCENTRATED FORM

Should start some spatial projects, to capture a large amount of energy somewhere near the source (near the Sun), energy which can be sent then to the Earth in a concentrated form (LASER, MASER, IRASER, etc).

The enormous energy emanating from the sun is spreading in all directions of the universe, and dilute with the distance.

On Earth no longer reach than a small amount from the energy emanated by the sun.

We try here (on the Earth) to capture a drop from a very small amount of energy, who came from Sun. And we also complain that the yield is low, and technological costs are high.

Installations which must do capturing the solar energy, could be installed over the Mercury.

From the Mercury, the concentrated energy will be transmitted directly focused on the Moon.

On the Moon, the energy will be conserved and forwarded to Earth in doses non-hazardous (with lower concentrations), using multi-channels microwaves.

CHAPTER I - PRESENTING OF AN ATOMIC MODEL AND SOME POSSIBLE APPLICATIONS IN LASER FIELD, H-hour

INTRODUCTION

This chapter presents, shortly, a new and original relation (20 & 20') who determines the radius with that, the electron is running around the nucleus of an atom [2].

In the picture number 1 one presents some electrons that are moving around the nucleus of an atom [1].

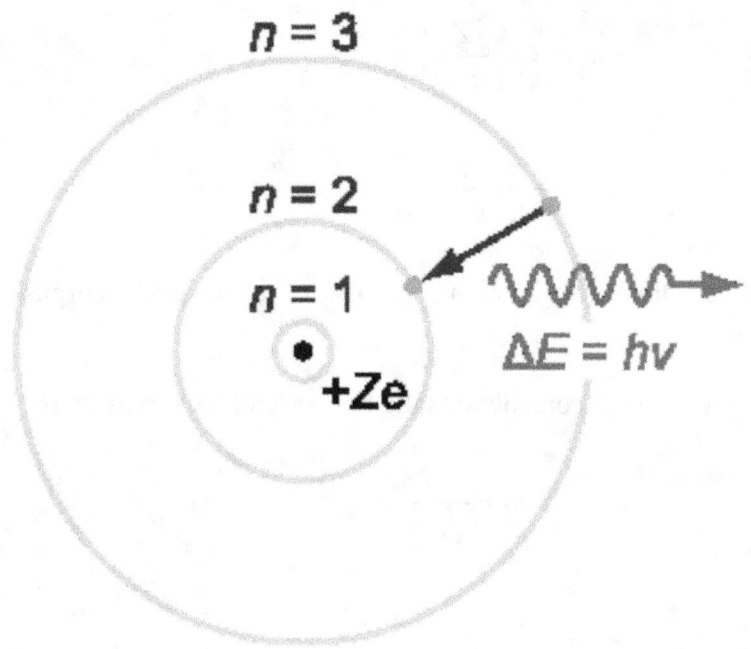

Fig. 1 *Electrons moving around the atomic nucleus;*

The atomic nucleus consists of nucleons (protons and neutrons)

One utilizes, two times the Lorenz relation (5), the Niels Bohr generalized equation (7), and a mass relation (4) which it was deduced from the kinematics energy relation written in two modes: classical (1) and coulombian (2). Equalizing the mass relation (4) with Lorenz relation (5) one obtains the form (6) which is a relation between the squared electron speed (v^2) and the radius (r).

The second relation (8), between v^2 and r, it was obtained by equalizing the mass of Bohr equation (7) and the mass of Lorenz relation (5).

In the system (8) – (6) eliminating the squared electron speed (v^2), it determines the radius r, with that the electron is moving around the atomic nucleus; see the relation (20).

For a Bohr energetically level (n=a constant value), one determines now two energetically below levels, which form an electronic layer.

The author realizes by this a new atomic model, or a new quantum theory, which explains the existence of electron-clouds without spin [1-2].

Writing the kinematics energy relation in two modes, classical (1) and coulombian (2) one determines the relation (3).

From the relation (3), determining explicit the mass of the electron, it obtains the form (4) [2].

$$E_C = \frac{1}{2} m \cdot v^2 \qquad (1)$$

$$E_C = \frac{1}{8} \frac{Z \cdot e^2}{\pi \cdot \varepsilon_0 \cdot r} \qquad (2)$$

$$m \cdot v^2 = \frac{1}{4} \frac{Z \cdot e^2}{\pi \cdot \varepsilon_0 \cdot r} \qquad (3)$$

$$m = \frac{Z \cdot e^2}{4 \cdot \pi \cdot \varepsilon_0 \cdot v^2 \cdot r} \qquad (4)$$

Now, we write the known relation Lorenz (5), for the mass of a corpuscle in function of the corpuscle squared speed.

With the relations (4) and (5) one obtains the first essential expression (6).

$$m = \frac{m_0 \cdot c}{\sqrt{c^2 - v^2}} \qquad (5)$$

$$\frac{m_0 \cdot c}{\sqrt{c^2 - v^2}} = \frac{Z \cdot e^2}{4 \cdot \pi \cdot \varepsilon_0 \cdot v^2 \cdot r} \qquad (6)$$

One utilizes now, the Niels Bohr generalized relation (7).

It uses for the second time the Lorenz relation (5) with the Bohr relation (7) and in this mode one obtains the second essential expression (8).

$$m = \frac{n^2 \cdot \varepsilon_0 \cdot h^2}{\pi \cdot r \cdot e^2 \cdot Z} \qquad (7)$$

$$\frac{m_0 \cdot c}{\sqrt{c^2 - v^2}} = \frac{n^2 \cdot \varepsilon_0 \cdot h^2}{\pi \cdot r \cdot e^2 \cdot Z} \qquad (8)$$

Now, one keeps just the two essential expressions (6 and 8). It writes (8) in the form (8').

$$\sqrt{c^2 - v^2} \cdot n^2 \cdot \varepsilon_0 \cdot h^2 = \pi \cdot r \cdot m_0 \cdot c \cdot e^2 \cdot Z \qquad (8')$$

Elevating the relationship (8') to the square, to explicit the squared electron speed, it obtains the form (9).

$$v^2 = \frac{(n^4 \cdot \varepsilon_0^2 \cdot h^4 - \pi^2 \cdot r^2 \cdot m_0^2 \cdot e^4 \cdot Z^2) \cdot c^2}{n^4 \cdot \varepsilon_0^2 \cdot h^4} \qquad (9)$$

The formula (9) can be put in the form (10), where the constant k takes the form (10').

$$v^2 = c^2 - k \cdot c^2 \cdot r^2 \qquad (10)$$

$$k = \frac{\pi^2 \cdot m_0^2 \cdot e^4 \cdot Z^2}{n^4 \cdot \varepsilon_0^2 \cdot h^4} \qquad (10')$$

Now one writes the essential relation (6) in the form (6').

$$4 \cdot m_0 \cdot c \cdot \pi \cdot \varepsilon_0 \cdot r \cdot v^2 = Z \cdot e^2 \cdot \sqrt{c^2 - v^2} \qquad (6')$$

Then, putting the relation (6') at the square, it obtains the formula (6'').

$$16 \cdot m_0^2 \cdot c^2 \cdot \pi^2 \cdot \varepsilon_0^2 \cdot r^2 \cdot v^4 = Z^2 \cdot e^4 \cdot (c^2 - v^2) \qquad (6'')$$

In the relation (6'') one introduce the squared velocity of the electron, taken from the expression (10) and one obtains the formula (11).

$$16 \cdot m_0^2 \cdot \pi^2 \cdot \varepsilon_0^2 \cdot (c^2 - k \cdot c^2 \cdot r^2)^2 = Z^2 \cdot e^4 \cdot k \qquad (11)$$

The (11) relationship can be arranged in the form (12).

$$(c^2 - k \cdot c^2 \cdot r^2)^2 = \frac{Z^2 \cdot e^4 \cdot k}{16 \cdot m_0^2 \cdot \pi^2 \cdot \varepsilon_0^2} \qquad (12)$$

One squares the relation (12) and it obtains the expression (13).

$$(c^2 - k \cdot c^2 \cdot r^2) = \pm \frac{Z \cdot e^2 \cdot \sqrt{k}}{4 \cdot m_0 \cdot \pi \cdot \varepsilon_0} \qquad (13)$$

The relation (13) can be arranged to the form (14).

$$k \cdot c^2 \cdot r^2 = c^2 \mp \frac{Z \cdot e^2 \cdot \sqrt{k}}{4 \cdot m_0 \cdot \pi \cdot \varepsilon_0} \qquad (14)$$

From relation (14) it explicit the squared electron radius and one obtains the relation (15).

$$r^2 = \frac{1}{k} \mp \frac{Z \cdot e^2}{4 \cdot m_0 \cdot \pi \cdot \varepsilon_0 \cdot \sqrt{k} \cdot c^2} \qquad (15)$$

Now, one exchange in the relation (15), the constant k with its expression (10') and it obtains the relation (16).

$$r^2 = \frac{n^4 \cdot \varepsilon_0^2 \cdot h^4}{\pi^2 \cdot m_0^2 \cdot e^4 \cdot Z^2} \mp \frac{n^2 \cdot h^2}{4 \cdot \pi^2 \cdot m_0^2 \cdot c^2} \qquad (16)$$

The expression (16) can be put in the form (17).

$$r^2 = \frac{n^4 \cdot \varepsilon_0^2 \cdot h^4}{\pi^2 \cdot m_0^2 \cdot e^4 \cdot Z^2} \cdot (1 \mp \frac{e^4 \cdot Z^2}{4 \cdot c^2 \cdot \varepsilon_0^2 \cdot h^2 \cdot n^2}) \qquad (17)$$

Extracting the square root of the expression (17), it obtains for the electron radius (r), the expression (18).

$$r = \pm \frac{n^2 \cdot \varepsilon_0 \cdot h^2}{\pi \cdot m_0 \cdot e^2 \cdot Z} \cdot \sqrt{1 \mp \frac{e^4 \cdot Z^2}{4 \cdot c^2 \cdot \varepsilon_0^2 \cdot h^2 \cdot n^2}} \qquad (18)$$

Physically there is only the positive solution (19).

$$r = + \frac{n^2 \cdot \varepsilon_0 \cdot h^2}{\pi \cdot m_0 \cdot e^2 \cdot Z} \cdot \sqrt{1 \mp \frac{e^4 \cdot Z^2}{4 \cdot c^2 \cdot \varepsilon_0^2 \cdot h^2 \cdot n^2}} \qquad (19)$$

The relation (19) is writing in final form (20) [3].

$$r = \frac{n^2 \cdot \varepsilon_0 \cdot h^2}{\pi \cdot m_0 \cdot e^2 \cdot Z} \cdot \sqrt{1 \mp \frac{e^4 \cdot {}^2}{4 \cdot c^2 \cdot \varepsilon_0^2 \cdot h^2 \cdot n^2}} \qquad (20)$$

The expression (20) it's not just a new theory for calculating the radius with that the electron is running around the nucleus of an atom, it is also a really new theory of an atomic model, or a new quantum theory.

For a value of the quantum number n (for a constant atomic number Z), we haven't just one energetically level (like in the Bohr model).

Now we can find two energetically below levels, which form an electronic layer, an electronic cloud. For example, for n=1, we have two sublevels (two below levels) [1-2].

USED NOTATIONS

The permissive constant (the permittivity): $\varepsilon_0 = 8.85418 \cdot 10^{-12} [\frac{C^2}{N \cdot m^2}]$;

The Planck constant: $h = 6.626 \cdot 10^{-34} [J \cdot s]$;

The rest mass of electron: $m_0 = 9.1091 \cdot 10^{-31} [kg]$;

The Pythagoras number: $\pi = 3.141592654$;

The electrical elementary load: $e = -1.6021 \cdot 10^{-19} [C]$;

The light speed in vacuum: $c = 2.997925 \cdot 10^{8} [\frac{m}{s}]$;

n=the principal quantum number (the Bohr quantum number);

Z=the number of protons from the atomic nucleus (the atomic number) [2].

DETERMINING THE TWO DIFFERENT ELECTRON SPEED VALUES

Relationship (6'') may be written in the form (6''') [2].

$$16 \cdot m_0^2 \cdot c^2 \cdot \pi^2 \cdot \varepsilon_0^2 \cdot r^2 \cdot v^4 + \\ + Z^2 \cdot e^4 \cdot v^2 - Z^2 \cdot e^4 \cdot c^2 = 0 \qquad (6''')$$

It can see easily that the relation (6''') represents a two degree equation in v^2.
One calculates v^2 with the formula (6^{IVa}).

$$v_{1,2}^2 = \frac{-Z^2 \cdot e^4 \pm \sqrt{Z^4 \cdot e^8 + 8^2 \cdot m_0^2 \cdot \pi^2 \cdot \varepsilon_0^2 \cdot c^4 \cdot Z^2 \cdot e^4 \cdot r^2}}{2 \cdot 16 \cdot m_0^2 \cdot c^2 \cdot \pi^2 \cdot \varepsilon_0^2 \cdot r^2} \qquad (6^{IVa})$$

Physically there is just the positive solution, and one keeps it for the relation (6^{IV}) (only the positive sign) [2].

$$v^2 = \frac{-Z^2 \cdot e^4 + \sqrt{Z^4 \cdot e^8 + 8^2 \cdot m_0^2 \cdot \pi^2 \cdot \varepsilon_0^2 \cdot c^4 \cdot Z^2 \cdot e^4 \cdot r^2}}{2 \cdot 16 \cdot m_0^2 \cdot c^2 \cdot \pi^2 \cdot \varepsilon_0^2 \cdot r^2} \qquad (6^{IV})$$

It can thinks that the relation (6^{IV}) gives only one solution for the electron squared speed (v^2), but really there is two solutions for this parameter, v^2, because the value of the squared radius (r^2) gives two physically solutions. It put the relation (6^{IV}) in the form (6^V) [2].

$$v_{1,2}^2 = \frac{-1 + \sqrt{1 + \dfrac{8^2 \cdot m_0^2 \cdot \pi^2 \cdot \varepsilon_0^2 \cdot c^2}{Z^2 \cdot e^4} \cdot c^2 \cdot r^2}}{\dfrac{1}{2} \cdot \dfrac{8^2 \cdot m_0^2 \cdot c^2 \cdot \pi^2 \cdot \varepsilon_0^2}{Z^2 \cdot e^4} \cdot r^2} \qquad (6^V)$$

The formula (6^V) can be written in the form (6^{VI}), where the constant k_1 takes the form (6^{VII}) [2].

$$v_{1,2}^2 = \frac{\sqrt{1 + k_1 \cdot c^2 \cdot r^2} - 1}{\dfrac{k_1}{2} \cdot r^2} \qquad (6^{VI})$$

$$k_1 = \frac{8^2 \cdot m_0^2 \cdot \pi^2 \cdot \varepsilon_0^2 \cdot c^2}{Z^2 \cdot e^4} \qquad (6^{VII})$$

Now one starts with relation (6^{VI}) who can be written in the form (21).

$$v^2 = \frac{2 \cdot c^2}{\sqrt{1 + k_1 \cdot c^2 \cdot r^2} + 1} \qquad (21)$$

One notes the radical with R (see the relation 22).

$$R = \sqrt{1 + k_1 \cdot c^2 \cdot r^2} \qquad (22)$$

In relation (22) one introduces for r^2 the expression (20) and it obtains the form (22').

$$R = \sqrt{1 + \frac{k_1 \cdot c^2}{k} \cdot (1 \mp \frac{2 \cdot \sqrt{k}}{c \cdot \sqrt{k_1}})} \qquad (22')$$

In relation (22') one exchanges the two constant k_1 and k with the two values from expressions (6^{VII}) respective (10') and it obtains for (22') the form (22'') [2].

$$R = \sqrt{1 + \frac{8^2 m_0^2 \cdot \pi^2 \cdot \varepsilon_0^2 \cdot c^4 \cdot n^4 \cdot \varepsilon_0^2 \cdot h^4}{Z^2 \cdot e^4 \cdot \pi^2 \cdot m_0^2 \cdot e^4 \cdot Z^2} \cdot (1 \mp \frac{2\pi \cdot m_0 \cdot e^4 \cdot Z^2}{8n^2 \cdot \varepsilon_0^2 \cdot h^2 \cdot c^2})} \qquad (22'')$$

One put the expression (22'') in the form (22''').

$$R = \sqrt{1 + \frac{8^2 \cdot \varepsilon_0^4 \cdot c^4 \cdot h^4 \cdot n^4}{e^8 \cdot Z^4}(1 \mp \frac{e^4 \cdot Z^2}{4\varepsilon_0^2 \cdot c^2 \cdot h^2 \cdot n^2})} \qquad (22''')$$

The expression (22''') will be written in the form (22^{IV}).

$$R = \sqrt{1 + \frac{8^2 \cdot \varepsilon_0^4 \cdot c^4 \cdot h^4 \cdot n^4}{e^8 \cdot Z^4} \mp \frac{2 \cdot 8 \cdot \varepsilon_0^2 \cdot c^2 \cdot h^2 \cdot n^2}{e^4 \cdot Z^2}} \qquad (22^{IV})$$

The expression (22^{IV}) can be restricted to the forms (22^V) and (22^{VI}).

$$R = \sqrt{\left(1 \mp \frac{8 \cdot \varepsilon_0^2 \cdot c^2 \cdot h^2 \cdot n^2}{e^4 \cdot Z^2}\right)^2} \qquad (22^V)$$

$$R = \left|1 \mp \frac{8 \cdot \varepsilon_0^2 \cdot c^2 \cdot h^2 \cdot n^2}{e^4 \cdot Z^2}\right| \qquad (22^{VI})$$

One notes with E the expression (23).

$$E = \frac{8 \cdot \varepsilon_0^2 \cdot c^2 \cdot h^2}{e^4} \cdot \frac{n^2}{Z^2} \qquad (23)$$

This expression must be evaluated.

$$E = \frac{8 \cdot 8.85418^2 \cdot 10^{-24} \cdot 2.997925^2 \cdot 10^{16}}{1.6021^4 \cdot 10^{-76}} \cdot$$
$$\cdot \frac{6.626^2 \cdot 10^{-68} \cdot n^2}{Z^2} = \frac{37564.06551 \cdot n^2}{Z^2} \qquad (23')$$

For Zmax=92, we have a minimum of expression E (23''):

$$E_{min} = 4.438098477 \cdot n^2 \qquad (23'')$$

It can see easily that Emin > 1:

$$E_{min} \succ 1 \qquad (24)$$

Now, one can write the expression (22^{VI}) in the forms (22^{VII}) a, and b:

$$R_1 = E - 1 \qquad (22^{VIIa})$$

$$R_2 = E + 1 \qquad (22^{VIIb})$$

Only now the expression (21) can be evaluated and reduced to two forms (21^{Ia}) and respective (21^{Ib}):

$$v_1^2 = \frac{2 \cdot c^2}{E - 1 + 1} \qquad (21^{Ia})$$

$$v_2^2 = \frac{2 \cdot c^2}{E + 1 + 1} \qquad (21^{Ib})$$

The two relations take the forms (21^{II}) a, and b:

$$v_1^2 = \frac{c^2}{\dfrac{E}{2}} \qquad (21^{IIa})$$

$$v_2^2 = \frac{c^2}{\dfrac{E}{2} + 1} \qquad (21^{IIb})$$

If one replaces E with its expression (23) it obtains for the electron speeds the relations (21^{III}) a, and b [2].

$$v_1^2 = \frac{e^4 \cdot Z^2}{4 \cdot \varepsilon_0^2 \cdot h^2 \cdot n^2} \qquad (21^{IIIa})$$

$$v_2^2 = \frac{c^2}{\dfrac{4 \cdot \varepsilon_0^2 \cdot c^2 \cdot h^2 \cdot ^2}{e^4 \cdot Z^2} + 1} \qquad (21^{IIIb})$$

DETERMINING THE MASSES AND THE ENERGY OF THE ATOMIC ELECTRON IN MOVEMENT

The exact squared speeds can be written in the forms (25, 26) [2].

$$r_- = r_1 \Rightarrow v_1^2 = \frac{e^4 \cdot Z^2 \cdot c^2}{4 \cdot \varepsilon_0^2 \cdot c^2 \cdot h^2 \cdot n^2} \tag{25}$$

$$r_+ = r_2 \Rightarrow v_2^2 = \frac{e^4 \cdot Z^2 \cdot c^2}{4 \cdot \varepsilon_0^2 \cdot c^2 \cdot h^2 \cdot n^2 + e^4 \cdot Z^2} \tag{26}$$

With these velocities one can write the two adequate masses (27), (28) [2].

$$r_- = r_1 \Rightarrow m_1 = \frac{m_0}{\sqrt{1 - \dfrac{e^4 \cdot Z^2}{4 \cdot \varepsilon_0^2 \cdot c^2 \cdot h^2 \cdot n^2}}} \tag{27}$$

$$r_+ = r_2 \Rightarrow m_2 = \frac{m_0}{\sqrt{1 - \dfrac{e^4 \cdot Z^2}{4 \cdot \varepsilon_0^2 \cdot c^2 \cdot h^2 \cdot n^2 + e^4 \cdot Z^2}}} \tag{28}$$

The total electron energy can be written in the forms (29) and (30) [2].

$$r_- = r_1 \Rightarrow W_1 = \frac{m_0 \cdot c^2}{\sqrt{1 - \frac{e^4 \cdot Z^2}{4 \cdot \varepsilon_0^2 \cdot c^2 \cdot h^2 \cdot n^2}}} \qquad (29)$$

$$r_+ = r_2 \Rightarrow W_2 = \frac{m_0 \cdot c^2}{\sqrt{1 - \frac{e^4 \cdot Z^2}{4 \cdot \varepsilon_0^2 \cdot c^2 \cdot h^2 \cdot n^2 + e^4 \cdot Z^2}}} \qquad (30)$$

The possible frequency of pumping, between the two near energetically below levels can be written in the form (31) [2].

$$v = \frac{W_1 - W_2}{h} = \frac{m_0 \cdot c^2}{h} \cdot \left[\frac{1}{\sqrt{1 - \frac{e^4 \cdot Z^2}{4 \cdot \varepsilon_0^2 \cdot c^2 \cdot h^2 \cdot n^2}}} - \frac{1}{\sqrt{1 - \frac{e^4 \cdot Z^2}{4 \cdot \varepsilon_0^2 \cdot c^2 \cdot h^2 \cdot n^2 + e^4 \cdot Z^2}}} \right] \qquad (31)$$

THE *POSSIBLE* LASER FREQUENCIES

In the table 1, one can see the possible LASER pumping frequencies (all in visible domain $4.34*10^{14} \div 6.97*10^{14}$ [Hz]), calculated for different principal quantum number n.

The possible L A S E R pumping frequencies — Table 1

n	Z	[zH]ν	Element	n	Z	[zH]ν	Element
2	15 =5.54942E14		P		78 =4.43344E+14		Pt
	22 =5.072E14		Ti		79 =4.66537E+14		Au
3	23 =6.0598E14		V		80 =4.90629E+14		Hg
	29 =4.8452E+14		Cu		81 =5.15642E+14		Tl
	30 =5.54942E+14		Zn		82 =5.41601E+14		Pb
4	31 =6.32782E+14		Ga		83 =5.68529E+14		Bi
	36 =4.71283E+14		Kr		84 =5.96449E+14		Po
	37 =5.25911E+14		Rb		85 =6.25386E+14		At
	38 =5.8516E+14		Sr		86 =6.55364E+14		Rn
5	39 =6.49284E+14		Y	11	87 =6.86408E+14		Fr
	43 =4.6261E+14		Tc		85 =4.41451E+14		At
	44 =5.072E+14		Ru		86 =4.6261E+14		Rn
	45 =5.54942E+14		Rh		87 =4.8452E+14		Fr
	46 =6.0598E+14		Pd		88 =5.072E+14		Ra
6	47 =6.60463E+14		Ag		89 =5.30668E+14		Ac
	50 =4.56488E+14		Sn		90 =5.54942E+14		Th
	51 =4.94145E+14		Sb		91 =5.8004E+14		Pa
	52 =5.34086E+14		Te		92 =6.0598E+14		U
	53 =5.76403E+14		I		93 =6.32782E+14		Np
	54 =6.21189E+14		Xe		94 =6.60463E+14		Pu
7	55 =6.68536E+14		Cs	12	95 =6.89044E+14		Am

	57 =4.51937E+14	La		92 =4.39854E+14	U
	58 =4.8452E+14	Ce		93 =4.59306E+14	Np
	59 =5.18835E+14	Pr		94 =4.79396E+14	Pu
	60 =5.54942E+14	Nd		95 =5.00139E+14	Am
	61 =5.92904E+14	Pm		96 =5.21548E+14	Cm
	62 =6.32782E+14	Sm		97 =5.43638E+14	Bk
8	63 =6.7464E+14	Eu		98 =5.66422E+14	Cf
	64 =4.48422E+14	Gd		99 =5.89916E+14	Es
	65 =4.77132E+14	Tb		100 =6.14134E+14	Fm
	66 =5.072E+14	Dy		101 =6.39091E+14	Md
	67 =5.38669E+14	Ho		102 =6.64801E+14	No
	68 =5.71581E14	Er	13	103 =6.9128E+14	Lw
	69 =6.0598E+14	Tm		99 =4.38489E+14	Es
	70 =6.4191E+14	Yb		100 =4.56488E+14	Fm
9	71 =6.79416E+14	Lu		101 =4.75037E+14	Md
	71 =4.45624E+14	Lu		102 =4.94145E+14	No
	72 =4.71283E+14	Hf		103 =5.13824E+14	Lr
	73 =4.98035E+14	Ta		104 =5.34086E+14	Rf
	74 =5.25911E+14	W	14	105 =5.54942E+14	Db
	75 =5.54942E+14	Re			
	76 =5.8516E+14	Os			
	77 =6.16596E+14	Ir			
	78 =6.49284E+14	Pt			
10	79 =6.83255E+14	Au			

THE LASER FREQUENCIES AND CONCLUSIONS

If the second speed value does not exist physically, we must calculate the new atomic model just for the new first value, with the next relations:

$$r = \frac{n^2 \cdot \varepsilon_0 \cdot h^2}{\pi \cdot m_0 \cdot e^2 \cdot Z} \cdot \sqrt{1 - \frac{e^4 \cdot Z^2}{4 \cdot c^2 \cdot \varepsilon_0^2 \cdot h^2 \cdot n^2}} \qquad (20')$$

$$v^2 = \frac{e^4 \cdot Z^2}{4 \cdot \varepsilon_0^2 \cdot h^2 \cdot n^2} \qquad (25')$$

$$m = \frac{m_0}{\sqrt{1 - \frac{e^4 \cdot Z^2}{4 \cdot \varepsilon_0^2 \cdot c^2 \cdot h^2 \cdot n^2}}} \qquad (27')$$

$$W = \frac{m_0 \cdot c^2}{\sqrt{1 - \frac{e^4 \cdot Z^2}{4 \cdot \varepsilon_0^2 \cdot c^2 \cdot h^2 \cdot n^2}}} \qquad (29')$$

$$\gamma = \frac{m_0 \cdot c^2}{h} \left(\frac{1}{\sqrt{1 - \frac{e^4 \cdot Z^2}{4 \cdot \varepsilon_0^2 \cdot c^2 \cdot h^2 \cdot n_1^2}}} - \frac{1}{\sqrt{1 - \frac{e^4 \cdot Z^2}{4 \cdot \varepsilon_0^2 \cdot c^2 \cdot h^2 \cdot n_2^2}}} \right) \qquad (31')$$

The pumping frequency required to achieve the transition of the electrons between two energetically levels can be written in the form (31').

In the table 2, one can see the LASER pumping frequencies.

All frequencies are outside visible area. One can make Ultraviolet Frequency-X ray LASER.

The bold value can be used to make a Rubin (Crystal) LASER.

The paper realizes a new atomic model and a new quantum theory (relation 20').

It determines as well the frequency of pumping for the transition between two energetically levels, with possible applications in LASER, MASER, IRASER industry (relation 31').

The pumping frequencies, between two nearer level									Table 2
Z	ν	El n_1-n_2	Z	ν	Element	Z	ν	Element	
1		H	2		He	3	2.22122E+16	Li 1-2	
4	3.95022E+16	Be 1-2	5	6.17499E+16	B 1-2	6	8.89688E+16	C 1-2	
7	1.21175E+17	N 1-2	8	1.58388E+17	O 1-2	9	2.00631E+17	F 1-2	
10	2.47929E+17	Ne 1-2	11	5.53738E+16	Na 2-3	12	6.59213E+16	Mg 2-3	
13	7.73939E+16	Al 2-3	14	8.97936E+16	Si 2-3	15	1.03123E+17	P 2-3	
16	1.17383E+17	S 2-3	17	1.32578E+17	Cl 2-3	18	1.48709E+17	Ar 2-3	
19	5.7866E+16	K 3-4	20	6.41348E+16	Ca 3-4	21	7.07288E+16	Sc 3-4	
22	7.76485E+16	Ti 3-4	23	8.48944E+16	V 3-4	24	9,24672E+16	Cr 3-4	
25	1.00368E+17	Mn 3-4	26	1.08596E+17	Fe 3-4	27	1.17153E+17	Co 3-4	
28	1.2604E+17	Ni 3-4	29	1.35258E+17	Cu 3-4	30	1.44806E+17	Zn 3-4	
31	1.54686E+17	Ga 3-4	32	1.64899E+17	Ge 3-4	33	1.75446E+17	As 3-4	
34	1.86327E+17	Se 3-4	35	1.97544E+17	Br 3-4	36	2.09097E+17	Kr 3-4	
37	1.01887E+17	Rb 4-5	38	1.07502E+17	Sr 4-5	39	1.1327E+17	Y 4-5	
40	1.19192E+17	Zr 4-5	41	1.25268E+17	Nb 4-5	42	1.31498E+17	Mo 4-5	
43	1.37882E+17	Tc 4-5	44	1.44421E+17	Ru 4-5	45	1.51116E+17	Rh 4-5	
46	1.57966E+17	Pd 4-5	47	1.64972E+17	Ag 4-5	48	1.72134E+17	Cd 4-5	
49	1.79453E+17	In 4-5	50	1.86928E+17	Sn 4-5	51	1.94561E+17	Sb 4-5	
52	2.02352E+17	Te 4-5	53	2.10301E+17	I 4-5	54	2.18408E+17	Xe 4-5	
55	1.22612E+17	Cs 5-6	56	1.2715E+17	Ba 5-6	57	1.31772E+17	La 5-6	
58	1.36479E+17	Ce 5-6	59	1.41271E+17	Pr 5-6	60	1.46147E+17	Nd 5-6	
61	1.51109E+17	Pm 5-6	62	1.56157E+17	Sm 5-6	63	1.6129E+17	Eu 5-6	
64	1.66508E+17	Gd 5-6	65	1.71813E+17	Tb 5-6	66	1.77203E+17	Dy 5-6	
67	1.8268E+17	Ho 5-6	68	1.88243E+17	Er 5-6	69	1.93893E+17	Tm 5-6	
70	1.9963E+17	Yb 5-6	71	2.05453E+17	Lu 5-6	72	2.11364E+17	Hf 5-6	
73	2.17362E+17	Ta 5-6	74	2.23448E+17	W 5-6	75	2.29621E+17	Re 5-6	
76	2.35883E+17	Os 5-6	77	2.42232E+17	Ir 5-6	78	2.4867E+17	Pt 5-6	
79	2.55197E+17	Au 5-6	80	2.61813E+17	Hg 5-6	81	2.68517E+17	Tl 5-6	
82	2.75311E+17	Pb 5-6	83	2.82195E+17	Bi 5-6	84	2.89168E+17	Po 5-6	
85	2.96231E+17	At 5-6	86	3.03385E+17	Rn 5-6	87	1.8618E+17	Fr 6-7	
88	1.90549E+17	Ra 6-7	89	1.94972E+17	Ac 6-7	90	1.99447E+17	Th 6-7	
91	2.03976E+17	Pa 6-7	92	2.08557E+17	U 6-7	93	2.13193E+17	Np 6-7	
94	2.17881E+17	Pu 6-7	95	2.22624E+17	Am 6-7	96	2.2742E+17	Cm 6-7	
97	2.3227E+17	Bk 6-7	98	2.37174E+17	Cf 6-7	99	2.42131E+17	Es 6-7	
100	2.47144E+17	Fm 6-7	101	2.5221E+17	Md 6-7	102	2.57331E+17	No 6-7	
103	2.62506E+17	Lr 6-7	104	2.67736E+17	Rf 6-7	105	2.73021E+17	Db 6-7	

THE RELATIONSHIPS

Determining the ray of an electron moving on an orbit around an atom

The main relationships 1 and 2 are written [2]:

$$\left.\begin{array}{l} \text{Kinetic energy } E_c = \frac{1}{2} \cdot m \cdot v^2 \\ \text{Coulomb form } E_C = \frac{1}{8} \cdot \frac{Z \cdot e^2}{\pi \cdot \varepsilon_0 \cdot r} \end{array}\right\} \Rightarrow m = \frac{Z \cdot e^2}{4 \cdot \pi \cdot \varepsilon_0 \cdot r \cdot v^2} \right\} \Rightarrow$$

$$\text{Lorentz relation } m = \frac{m_0 \cdot c}{\sqrt{c^2 - v^2}}$$

$$\Rightarrow \frac{m_0 \cdot c}{\sqrt{c^2 - v^2}} = \frac{Z \cdot e^2}{4 \cdot \pi \cdot \varepsilon_0 \cdot r \cdot v^2} \Rightarrow \begin{cases} l \cdot r \cdot c \cdot v^2 = \sqrt{c^2 - v^2} \\ \text{with } l = \frac{4 \cdot \pi \cdot m_0 \cdot \varepsilon_0}{Z \cdot e^2} \end{cases}$$

(1)

$$\left.\begin{array}{l} \text{Niels Bohr relation } m = \frac{\varepsilon_0 \cdot h^2 \cdot n^2}{\pi \cdot e^2 \cdot Z \cdot r} \\ \text{Lorentz relation } m = \frac{m_0 \cdot c}{\sqrt{c^2 - v^2}} \end{array}\right\} \Rightarrow \frac{m_0 \cdot c}{\sqrt{c^2 - v^2}} = \frac{\varepsilon_0 \cdot h^2 \cdot n^2}{\pi \cdot e^2 \cdot Z \cdot r} \Rightarrow$$

$$\Rightarrow \frac{\pi \cdot m_0 \cdot e^2 \cdot Z}{\varepsilon_0 \cdot h^2 \cdot n^2} \cdot r \cdot c = \sqrt{c^2 - v^2} \Rightarrow \begin{cases} k \cdot r \cdot c = \sqrt{c^2 - v^2} \\ \text{with } k = \frac{\pi \cdot m_0 \cdot e^2 \cdot Z}{\varepsilon_0 \cdot h^2 \cdot n^2} \end{cases}$$

(2)

It put the relationship 2 at the square and we obtain the formula 3.

$$v^2 = c^2 - k^2 \cdot r^2 \cdot c^2 \tag{3}$$

3 is inserted in the relationship 1 and we obtain the relations 4.

$$\begin{cases} l \cdot r \cdot c \cdot (c^2 - k^2 \cdot r^2 \cdot c^2) = \sqrt{c^2 - c^2 + k^2 \cdot r^2 \cdot c^2} \Rightarrow l \cdot r \cdot c \cdot c^2 \cdot (1 - k^2 \cdot r^2) = \sqrt{k^2 \cdot r^2 \cdot c^2} \\ l \cdot r \cdot c \cdot c^2 \cdot (1 - k^2 \cdot r^2) = \pm r \cdot c \cdot k \Rightarrow l \cdot c^2 \cdot (1 - k^2 \cdot r^2) = \pm k \Rightarrow 1 - k^2 \cdot r^2 = \pm \dfrac{k}{l \cdot c^2} \Rightarrow \\ r^2 = \dfrac{1}{k^2} \cdot \left(1 \mp \dfrac{k}{l \cdot c^2}\right) \Rightarrow r = \pm \dfrac{1}{k} \sqrt{1 \mp \dfrac{k}{l \cdot c^2}} \Rightarrow r = \dfrac{1}{k} \sqrt{1 \mp \dfrac{k}{l \cdot c^2}} \Rightarrow \\ r = \dfrac{\varepsilon_0 \cdot h^2 \cdot n^2}{\pi \cdot m_0 \cdot e^2 \cdot Z} \cdot \sqrt{1 \mp \dfrac{\pi \cdot m_0 \cdot e^2 \cdot Z \cdot e^2 \cdot Z}{n^2 \cdot \varepsilon_0 \cdot h^2 \cdot 4 \cdot \pi \cdot m_0 \cdot \varepsilon_0 \cdot c^2}} \Rightarrow \\ r = \dfrac{\varepsilon_0 \cdot h^2 \cdot n^2}{\pi \cdot m_0 \cdot e^2 \cdot Z} \cdot \sqrt{1 \mp \dfrac{e^4 \cdot Z^2}{4 \cdot \varepsilon_0^2 \cdot h^2 \cdot n^2 \cdot c^2}} \end{cases}$$

(4)

The final form (in 4) determines the ray of an electron running on an orbit around an atom. We have two r values at a single principal quantum number, n. It obtains a new and doubled relationship [2].

Determining the velocities of an electron which is running around an atom

From relationship 1 it obtains the speed of an electron to the square. We determine relationships numbered with 5.

$$\begin{cases} v^2 = \dfrac{2 \cdot c^2}{1 + \sqrt{1 + 4 \cdot c^4 \cdot r^2 \cdot l^2}} \Rightarrow v^2 = \dfrac{2 \cdot c^2}{1 + R} \quad \text{with} \quad R = \sqrt{1 + 4 \cdot c^4 \cdot r^2 \cdot l^2} \\ R = \sqrt{1 + 4 \cdot c^4 \cdot r^2 \cdot l^2} = \sqrt{1 + \dfrac{4 \cdot c^4 \cdot l^2}{k^2} \mp 2 \cdot \dfrac{2 \cdot c^2 \cdot l}{k}} = \sqrt{\left(1 \mp 2 \cdot \dfrac{c^2 \cdot l}{k}\right)^2} = \\ = \left|1 \mp 2 \cdot \dfrac{c^2 \cdot l}{k}\right| = \begin{cases} \dfrac{2 \cdot c^2 \cdot l}{k} - 1 = \dfrac{8 \cdot \varepsilon_0^2 \cdot h^2 \cdot n^2 \cdot c^2}{e^4 \cdot Z^2} - 1 \\ \dfrac{2 \cdot c^2 \cdot l}{k} + 1 = \dfrac{8 \cdot \varepsilon_0^2 \cdot h^2 \cdot n^2 \cdot c^2}{e^4 \cdot Z^2} + 1 \end{cases} \quad \text{with} \quad E = \dfrac{2 \cdot c^2 \cdot l}{k} > 1 \\ v_-^2 = \dfrac{2 \cdot c^2}{1 + \dfrac{8 \cdot \varepsilon_0^2 \cdot h^2 \cdot n^2 \cdot c^2}{e^4 \cdot Z^2} - 1} = \dfrac{2 \cdot c^2}{\dfrac{8 \cdot \varepsilon_0^2 \cdot h^2 \cdot n^2 \cdot c^2}{e^4 \cdot Z^2}} = \dfrac{c^2}{\dfrac{4 \cdot \varepsilon_0^2 \cdot h^2 \cdot n^2 \cdot c^2}{e^4 \cdot Z^2}} = \dfrac{k}{l} \\ v_+^2 = \dfrac{2 \cdot c^2}{1 + \dfrac{8 \cdot \varepsilon_0^2 \cdot h^2 \cdot n^2 \cdot c^2}{e^4 \cdot Z^2} + 1} = \dfrac{2 \cdot c^2}{\dfrac{8 \cdot \varepsilon_0^2 \cdot h^2 \cdot n^2 \cdot c^2}{e^4 \cdot Z^2} + 2} = \dfrac{c^2}{\dfrac{4 \cdot \varepsilon_0^2 \cdot h^2 \cdot n^2 \cdot c^2}{e^4 \cdot Z^2} + 1} = \dfrac{kc^2}{lc^2 + k} \end{cases}$$

(5)

Determining the mass of the electron in movement

When the speeds are known is simple to find quickly the masses values (forms 6).

$$m_- = \frac{m_0}{\sqrt{1 - \dfrac{1}{\dfrac{4 \cdot \varepsilon_0^2 \cdot h^2 \cdot n^2 \cdot c^2}{e^4 \cdot Z^2}}}} \qquad m_+ = \frac{m_0}{\sqrt{1 - \dfrac{1}{\dfrac{4 \cdot \varepsilon_0^2 \cdot h^2 \cdot n^2 \cdot c^2}{e^4 \cdot Z^2} + 1}}} \qquad (6)$$

Determining the energy of the electron in movement

To determine the energy of an electron in movement, it multiplies the mass of an electron with the squared speed of light (using the Einstein relation), (forms 7).

$$W_- = \frac{m_0 \cdot c^2}{\sqrt{1 - \dfrac{1}{\dfrac{4 \cdot \varepsilon_0^2 \cdot h^2 \cdot n^2 \cdot c^2}{e^4 \cdot Z^2}}}} \qquad W_+ = \frac{m_0 \cdot c^2}{\sqrt{1 - \dfrac{1}{\dfrac{4 \cdot \varepsilon_0^2 \cdot h^2 \cdot n^2 \cdot c^2}{e^4 \cdot Z^2} + 1}}} \qquad (7)$$

Determining the frequencies of pumping

Finally, we can write the frequency of pumping between the two energetic sub levels, adjacent (see the form 8).

$$\nu = \frac{W_1 - W_2}{h} = \frac{m_0 \cdot c^2}{h} \cdot \left(\frac{1}{\sqrt{1 - \dfrac{1}{\dfrac{4 \cdot \varepsilon_0^2 \cdot h^2 \cdot c^2 \cdot n^2}{e^4 \cdot Z^2}}}} - \frac{1}{\sqrt{1 - \dfrac{1}{\dfrac{4 \cdot \varepsilon_0^2 \cdot h^2 \cdot c^2 \cdot n^2}{e^4 \cdot Z^2} + 1}}} \right) \qquad (8)$$

Notes utilized (used notations) (forms 9)

$$\begin{cases} \text{The permissive constant (the permittivity)}: \quad \varepsilon_0 = 8.85418 \cdot 10^{-12} \; [\frac{C^2}{N \cdot m^2}] \\[4pt] \text{The Planck constant}: \quad h = 6.626 \cdot 10^{-34} \; [J \cdot s] \\[4pt] \text{The rest mass of electron}: \quad m_0 = 9.1091 \cdot 10^{-31} \; [kg] \\[4pt] \text{The Pythagora's number}: \quad \pi = 3.141592654 \\[4pt] \text{The electrical elementary load}: \quad e = -1.6021 \cdot 10^{-19} \; [C] \\[4pt] \text{The light speed in vacuum}: \quad c = 2.997925 \cdot 10^8 [\frac{m}{s}] \\[4pt] n = \text{the principal quantum number (the Bohr quantum number)} \\ Z = \text{the number of protons from the atomic nucleus (the atomic number)} \end{cases} \quad (9)$$

Table 3: The LASER frequencies of pumping (n=2-5)

n	Z	ע[zH]	Element
2	15	=5.54942E14	P
	22	=5.072E14	Ti
3	23	=6.0598E14	V
	29	=4.8452E+14	Cu
	30	=5.54942E+14	Zn
4	31	=6.32782E+14	Ga
	37	=5.25911E+14	Rb
	38	=5.8516E+14	Sr
5	39	=6.49284E+14	Y

CONCLUSIONS

All frequencies, calculated in the table 1, are outside of the visible domain ($4.34*10^{14} \div 6.97*10^{14}$ [Hz]).

Only the atmospheric elements, N and O, are located near the visible frequencies when n=1.

The bold value can be used to make a Rubin (Crystal) LASER.

For n=2-5 there are nine values indicated to make a LASER in the visible domain (see the table 2).

The substance is structured in this mode, that, we can obtain more energy, if one can penetrate it deeply. In this mode, we can check and extract, small portions of energy, but the total obtained energy will be bigger.

The atomic electrons are coupled. The transition between the two coupled electrons can give us more energy, in small portions.

First, we can make a stronger "Electromagnetic Amplification by the Stimulated Emission of Radiation" LASER (MASER), by pumping the energy between two sub levels, adjacent.

This paper briefly describes how to determine the relationships by which it calculates the ray of an electron moving on an orbit around an atom.

Now, it's the time to correct the length of the r radius (see the Cap. 5. (4)→(12)).

CORRECTING THE LENGTH OF THE R RADIUS

The main expression (2) can be written in the form (10).

$$r = \frac{1}{k} \cdot \sqrt{1 - \frac{v^2}{c^2}} \qquad (10)$$

The velocities have the forms (11), known.

$$\begin{cases} v_-^2 = \dfrac{k \cdot c^2}{l \cdot c^2} = \dfrac{k}{l} \\ v_+^2 = \dfrac{k \cdot c^2}{l \cdot c^2 + k} \end{cases} \qquad (11)$$

With the relations (11) the expression (10) takes the forms (12).

$$\begin{cases} r_- = \dfrac{1}{k} \cdot \sqrt{1 - \dfrac{k}{l \cdot c^2}} = \dfrac{\varepsilon_0 \cdot h^2 \cdot n^2}{\pi \cdot m_0 \cdot e^2 \cdot Z} \cdot \sqrt{1 - \dfrac{e^4 \cdot Z^2}{4 \cdot \varepsilon_0^2 \cdot h^2 \cdot n^2 \cdot c^2}} \\ r_+ = \dfrac{1}{k} \cdot \sqrt{1 - \dfrac{k}{l \cdot c^2 + k}} = \dfrac{\varepsilon_0 \cdot h^2 \cdot n^2}{\pi \cdot m_0 \cdot e^2 \cdot Z} \cdot \sqrt{1 - \dfrac{e^4 \cdot Z^2}{4 \cdot \varepsilon_0^2 \cdot h^2 \cdot n^2 \cdot c^2 + e^4 \cdot Z^2}} \end{cases} \qquad (12)$$

The values imposed by relations 12 are probably the real physical values, because the main relations 1 and 2 are verified in the same time by the relationships 12.

The velocities, masses, energies and frequency of pumping have not changed (see a recap in cap. 6, relations 13-16).

RECAP

$$\begin{cases} r_- = \dfrac{1}{k} \cdot \sqrt{1 - \dfrac{k}{l \cdot c^2}} = \dfrac{\varepsilon_0 \cdot h^2 \cdot n^2}{\pi \cdot m_0 \cdot e^2 \cdot Z} \cdot \sqrt{1 - \dfrac{e^4 \cdot Z^2}{4 \cdot \varepsilon_0^2 \cdot h^2 \cdot n^2 \cdot c^2}} \\ r_+ = \dfrac{1}{k} \cdot \sqrt{1 - \dfrac{k}{l \cdot c^2 + k}} = \dfrac{\varepsilon_0 \cdot h^2 \cdot n^2}{\pi \cdot m_0 \cdot e^2 \cdot Z} \cdot \sqrt{1 - \dfrac{e^4 \cdot Z^2}{4 \cdot \varepsilon_0^2 \cdot h^2 \cdot n^2 \cdot c^2 + e^4 \cdot Z^2}} \end{cases} \quad (13)$$

$$\begin{cases} v_-^2 = \dfrac{2 \cdot c^2}{1 + \dfrac{8 \cdot \varepsilon_0^2 \cdot h^2 \cdot n^2 \cdot c^2}{e^4 \cdot Z^2} - 1} = \dfrac{2 \cdot c^2}{\dfrac{8 \cdot \varepsilon_0^2 \cdot h^2 \cdot n^2 \cdot c^2}{e^4 \cdot Z^2}} = \dfrac{c^2}{\dfrac{4 \cdot \varepsilon_0^2 \cdot h^2 \cdot n^2 \cdot c^2}{e^4 \cdot Z^2}} = \dfrac{k}{l} \\ v_+^2 = \dfrac{2 \cdot c^2}{1 + \dfrac{8 \cdot \varepsilon_0^2 \cdot h^2 \cdot n^2 \cdot c^2}{e^4 \cdot Z^2} + 1} = \dfrac{2 \cdot c^2}{\dfrac{8 \cdot \varepsilon_0^2 \cdot h^2 \cdot n^2 \cdot c^2}{e^4 \cdot Z^2} + 2} = \dfrac{c^2}{\dfrac{4 \cdot \varepsilon_0^2 \cdot h^2 \cdot n^2 \cdot c^2}{e^4 \cdot Z^2} + 1} = \dfrac{k \cdot c^2}{l \cdot c^2 + k} \end{cases} \quad (14)$$

$$m_- = \dfrac{m_0}{\sqrt{1 - \dfrac{1}{\dfrac{4 \cdot \varepsilon_0^2 \cdot h^2 \cdot n^2 \cdot c^2}{e^4 \cdot Z^2}}}} \qquad m_+ = \dfrac{m_0}{\sqrt{1 - \dfrac{1}{\dfrac{4 \cdot \varepsilon_0^2 \cdot h^2 \cdot n^2 \cdot c^2}{e^4 \cdot Z^2} + 1}}} \quad (15)$$

$$W_- = \dfrac{m_0 \cdot c^2}{\sqrt{1 - \dfrac{1}{\dfrac{4 \cdot \varepsilon_0^2 \cdot h^2 \cdot n^2 \cdot c^2}{e^4 \cdot Z^2}}}} \qquad W_+ = \dfrac{m_0 \cdot c^2}{\sqrt{1 - \dfrac{1}{\dfrac{4 \cdot \varepsilon_0^2 \cdot h^2 \cdot n^2 \cdot c^2}{e^4 \cdot Z^2} + 1}}} \quad (16)$$

H-hour

With the help of powerful lasers one can create a dense and highly ionized plasma. We need a highly ionized dense plasma to achieve nuclear fusion (cold or hot).

Since 1989, it talks about achieving nuclear fusion hot and cold. Another two decades have passed and humanity still does not benefit from nuclear fusion energy. What actually happens? Is it an unattainable myth? It was also circulated by the media that has been achieved nuclear fusion heat. Since 1989 there are all sorts of scientists with all kinds of crafted devices, which declare that they can produce nuclear power obtained by cold fusion (using cold plasma). May be that these devices works, but their yield is probably too small, or at an enlarged scale these give not the expected results. This is the real reason why we can't use yet the survival fuel (the deuterium).

Unfortunately today the dominant processes that produce energy are combustion (reaction) chemical combination of carbon with oxygen. Thermal energy released from such reactions is conventionally valued at about 7000 calories per gram.

Only the early 20th century physicists have succeeded in producing, other energy than by traditional methods. Energy release per unit mass was enormous compared with that obtained by conventional procedures. The Kilowatt based on nuclear fission of uranium nuclei has today a significant share in global energy balance. Unfortunately, the nuclear power plants burn the fuel uranium, already considered conventional and on extinct.

The current nuclear power is considered a transition way, to the energy thermonuclear, based on fusion of light nuclei.

The main particularity of synthesis reaction (fusion) is the high prevalence of the used fuel (primary), deuterium. It can be obtained relatively simply from ordinary water.

Deuterium was extracted from water for the first time by Harold Urey in 1931. Even at that time, small linear electrostatic accelerators, have indicated that D-D reaction (fusion of two deuterium nuclei) is exothermic.

Today we know that not only the first isotope of hydrogen (deuterium) produces fusion energy, but and the second (heavy) isotope of hydrogen (tritium) can produce energy by nuclear fusion.

The first reaction is possible between two nuclei of deuterium, from which can be obtained, either a tritium nucleus plus a proton and energy, or an isotope of helium with a neutron and energy.

$$_1^2D + {_1^2}D \rightarrow \begin{cases} _1^3T + 1MeV + {_1^1}H + 3MeV = {_1^3}T + {_1^1}H + 4MeV \\ _2^3He + 0.8MeV + {^1}n + 2.5MeV = {_2^3}He + {^1}n + 3.3MeV \end{cases}$$

Observations: a deuterium nucleus has a proton and a neutron; a tritium nucleus has a proton and two neutrons.

Fusion can occur between a nucleus of deuterium and one of tritium.

$$^2_1D + ^3_1T \rightarrow ^4_2He + 3.5 MeV + ^1n + 14 MeV = ^4_2He + ^1n + 17.5 MeV$$

Another fusion reaction can be produced between a nucleus of deuterium and an isotope of helium.

$$^2_1D + ^3_2He \rightarrow ^4_2He + 3.7 MeV + ^1_1H + 14.7 MeV = ^4_2He + ^1_1H + 18.4 MeV$$

For these reactions to occur, should that the deuterium nuclei have enough kinetic energy to overcome the electrostatic forces of rejection due to the positive tasks of protons in the nuclei.

For deuterium, for average kinetic energy are required tens of keV.

For 1 keV are needed about 10 million degrees temperature. For this reason hot fusion requires a temperature of hundreds of millions of degrees.
The huge temperature is done with high power lasers acting hot plasma.
Electromagnetic fields are arranged so that it can maintain hot plasma.

The best results were obtained with the Tokamak-type installations (see the Figure below).

ITER: the world's largest Tokamak

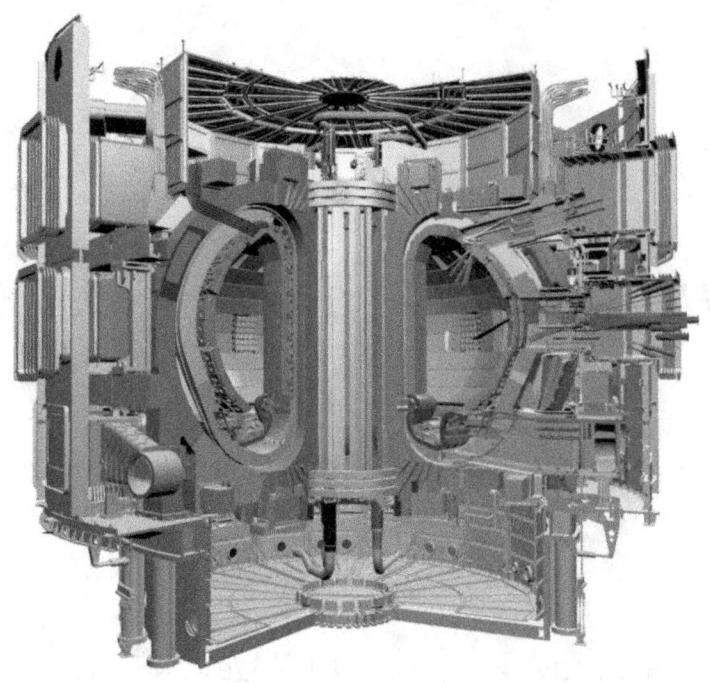

ITER is based on the 'tokamak' concept of magnetic confinement, in which the plasma is contained in a doughnut-shaped vacuum vessel. The fuel—a mixture of deuterium and tritium, two isotopes of hydrogen—is heated to temperatures in excess of 150 million°C, forming a hot plasma. Strong magnetic fields are used to keep the plasma away from the walls; these are produced by superconducting coils surrounding the vessel, and by an electrical current driven through the plasma.

Deuterium fuel is delivered in heavy water, D_2O.
Tritium is obtained in the laboratory by the following reaction.

$$^6_3Li + ^1 n \rightarrow ^3_1 T + ^4_2 He + 4.6 MeV$$

Lithium, the third element in Mendeleev's table, is found in nature in sufficient quantities.
The accelerated neutrons which produce the last presented reaction with lithium, appear from the second and the third presented reaction.
Raw materials for fusion are deuterium and lithium.

All fusion reactions shown produce finally energy and He. He is a (gas) inert element. Because of this, fusion reaction is clean, and far superior to nuclear fission.

Hot fusion works with very high temperatures.

In cold fusion, it must accelerate the deuterium nucleus, in linear or circular accelerators. Final energy of accelerated deuterium nuclei should be well calibrated for a positive final yield of fusion reactions (more mergers, than fission).
Electromagnetic fields which maintain the plasma (cold and especially the warm), should be and constrictors (especially at cold fusion), for to press, and more close together the nuclei.

The potential energy with that two protons reject each other, be calculated approximately with the following relationship.

$$U \equiv E_p = \frac{1}{4 \cdot \pi \cdot \varepsilon_0} \cdot \frac{q_1 \cdot q_2}{d_{12}} = \frac{1}{4 \cdot \pi \cdot 8.8541853 \cdot 10^{-12}} \cdot \frac{(1.602 \cdot 10^{-19})^2}{4 \cdot 10^{-15}} =$$

$$= 5.7664 \cdot 10^{-14} [J] = 5.7664 \cdot 10^{-14} \cdot 6.242 \cdot 10^{18} [eV] = 3.599 \cdot 10^5 [eV] =$$

$$= 360 [keV]$$

At a keV is necessary a temperature of 10 million ^0C.
At 360 keV is necessary a temperature of 3600 million ^0C.

In hot fusion it need a temperature of 3600 million degrees.

Without a minimum of 3000 million degrees we can't make the hot fusion reaction, to obtain the nuclear power.

Today we have just 150 million degrees made.

Fusion reactions of light elements power the stars and produce virtually all elements in a process called nucleosynthesis.
The fusion of lighter elements in stars releases energy (and the mass that always accompanies it).
For example, in the fusion of two hydrogen nuclei to form helium, seven-tenths of 1 percent of the mass is carried away from the system in the form of kinetic energy or other forms of energy (such as electromagnetic radiation). However, the production of elements heavier than iron, absorbs energy.
Research into controlled fusion, with the aim of producing fusion power for the production of electricity, has been conducted for over 60 years. It has been accompanied by extreme scientific and technological difficulties, but has resulted in progress.
At present, controlled fusion reactions have been unable to produce break-even (self-sustaining) controlled fusion reactions.
Workable designs for a reactor that theoretically will deliver ten times more fusion energy than the amount needed to heat up plasma to required temperatures (see ITER) were originally scheduled to be operational in 2018, however this has been delayed and a new date has not been stated.
It takes considerable energy to force nuclei to fuse, even those of the lightest element, hydrogen.
This is because all nuclei have a positive charge (due to their protons), and as like charges repel, nuclei strongly resist being put too close together.
Accelerated to high speeds (that is, heated to thermonuclear temperatures), they can overcome this electrostatic repulsion and get close enough for the attractive nuclear force to be sufficiently strong to achieve fusion.

The fusion of lighter nuclei, which creates a heavier nucleus and often a free neutron or proton, generally releases more energy than it takes to force the nuclei together; this is an exothermic process that can produce self-sustaining reactions.

The US National Ignition Facility, which uses laser-driven inertial confinement fusion, is thought to be capable of break-even fusion.

Energy released in most nuclear reactions is much larger than in chemical reactions, because the binding energy that holds a nucleus together is far greater than the energy that holds electrons to a nucleus.

For example, the ionization energy gained by adding an electron to a hydrogen nucleus is 13.6 eV—less than one-millionth of the 17 MeV released in the deuterium–tritium (D–T) reaction shown in the diagram to the right.

Fusion reactions have an energy density many times greater than nuclear fission; the reactions produce far greater energies per unit of mass even though individual fission reactions are generally much more energetic than individual fusion ones, which are themselves millions of times more energetic than chemical reactions.

Only direct conversion of mass into energy, such as that caused by the annihilation collision of matter and antimatter, is more energetic per unit of mass than nuclear fusion.

A substantial energy barrier of electrostatic forces must be overcome before fusion can occur. At large distances two naked nuclei repel one another because of the repulsive electrostatic force between their positively charged protons.

If two nuclei can be brought close enough together, however, the electrostatic repulsion can be overcome by the attractive nuclear force, which is stronger at close distances.

When a nucleon such as a proton or neutron is added to a nucleus, the nuclear force attracts it to other nucleons, but primarily to its immediate neighbours due to the short range of the force.

The nucleons in the interior of a nucleus have more neighboring nucleons than those on the surface.

Since smaller nuclei have a larger surface area-to-volume ratio, the binding energy per nucleon due to the nuclear force generally increases with the size of the nucleus but approaches a limiting value corresponding to that of a nucleus with a diameter of about four nucleons.

It is important to keep in mind that the above picture is a toy model because nucleons are quantum objects, and so, for example, since two neutrons in a nucleus are identical to each other, distinguishing one from the other, such as which one is in the interior and which is on the surface, is in fact meaningless, and the inclusion of quantum mechanics is necessary for proper calculations.

The electrostatic force, on the other hand, is an inverse-square force, so a proton added to a nucleus will feel an electrostatic repulsion from all the other protons in the nucleus.

The electrostatic energy per nucleon due to the electrostatic force thus increases without limit as nuclei get larger.

In hot fusion it need a temperature of 3600 million degrees.
Without a minimum of 3000 million degrees we can't make the hot fusion reaction, to obtain the nuclear power.

Today we have just 150 million degrees made.

To replace the lack of necessary temperature, it uses various tricks.

In cold fusion one must accelerate the deuterium nuclei at an energy of 390 [keV], and then collide them with the cold fusion fuel (heavy water and lithium).

Cold Nuclear Fusion

Because obtaining the necessary huge temperature for hot fusion is still difficult, it is time to focus us on cold nuclear fusion.
We need to bomb the fuel with accelerated deuterium nuclei.
The fuel will be made from heavy water and lithium.
The optimal proportion of lithium will be tested.
It would be preferable to keep fuel in the plasma state.
Between deuterium and tritium the smallest radius is the radius of deuterium nucleus.

$Deuterium \quad A = 2 \quad A^{1/3} = 1.259921 \Rightarrow R_D = 1.8268855223476 \cdot 10^{-15} [m]$

$Tritium \quad A = 3 \quad A^{1/3} = 1.44224957 \Rightarrow R_T = 2.0912618769457 \cdot 10^{-15} [m]$

We calculate the minimum distance between two particles which meet together. This is just the diameter of a deuterium nucleus, d_{12D}.

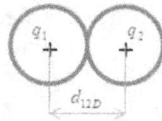

$$d_{12D} = 2 \cdot R_D = 2 \cdot 1.8268855223476 \cdot 10^{-15} [m] =$$
$$= 3.6537710446952 \cdot 10^{-15} [m] =$$
$$\approx 3.653771 \cdot 10^{-15} [m]$$

The deuterium nuclei which will bomb the nuclear fuel, will be accelerated with the (least) energy which reject the two neighboring deuterium nuclei (see the below relationship).

$$U \equiv E_p = \frac{1}{4\cdot\pi\cdot\varepsilon_0}\cdot\frac{q_1\cdot q_2}{d_{12D}} = \frac{1}{4\cdot\pi\cdot 8.8541853\cdot 10^{-12}}\cdot\frac{(1.602\cdot 10^{-19})^2}{3.653771\cdot 10^{-15}} =$$

$$= 6.3128464855\cdot 10^{-14}[J] = 6.3128464855\cdot 10^{-14}\cdot 6.242\cdot 10^{18}[eV] =$$

$$= 3.94\cdot 10^5[eV] = 3.94\cdot 10^2[keV] = 394[keV]$$

Aneutronic fusion

Aneutronic fusion is any form of fusion power where neutrons carry no more than 1% of the total released energy. The most-studied fusion reactions release up to 80% of their energy in neutrons.

Successful aneutronic fusion would greatly reduce problems associated with neutron radiation such as ionizing damage, neutron activation, and requirements for biological shielding, remote handling, and safety.

Some proponents also see a potential for dramatic cost reductions by converting energy directly to electricity. However, the conditions required to harness aneutronic fusion are much more extreme than those required for the conventional deuterium–tritium (DT) fuel cycle.

Candidate aneutronic reactions

There are a few fusion reactions that have no neutrons as products on any of their branches. Those with the largest cross sections are these:

$D + {}^3He \rightarrow {}^4He\ (3.6\ MeV) + p\ (14.7\ MeV)$

$D + {}^6Li \rightarrow 2\ {}^4He + 22.4\ MeV$

$p + {}^6Li \rightarrow {}^4He\ (1.7\ MeV) + {}^3He\ (2.3\ MeV)$

${}^3He + {}^6Li \rightarrow 2\ {}^4He + p + 16.9\ MeV$

${}^3He + {}^3He \rightarrow {}^4He + 2p + 12.86\ MeV$

$p + {}^7Li \rightarrow 2\ {}^4He + 17.2\ MeV$

$p + {}^{11}B \rightarrow 3\ {}^4He + 8.7\ MeV$

$p + {}^{15}N \rightarrow {}^{12}C + {}^4He + 5.0\ MeV$

The two of these which use deuterium as a fuel produce some neutrons with D–D side reactions.

Although these can be minimized by running hot and deuterium-lean, the fraction of energy released as neutrons will probably be several percent, so that these fuel cycles, although neutron-poor, do not qualify as aneutronic according to the 1% threshold.

The next two reactions' rates (involving p, ^{3}He, and ^{6}Li) are not particularly high in a thermal plasma.

When treated as a chain, however, they offer the possibility of enhanced reactivity due to a non-thermal distribution.

The product ^{3}He from the first reaction could participate in the second reaction before thermalizing, and the product p from the second reaction could participate in the first reaction before thermalizing.

Unfortunately, detailed analyses do not show sufficient reactivity enhancement to overcome the inherently low cross section.

The pure ^{3}He reaction suffers from a fuel-availability problem. ^{3}He occurs in only minuscule amounts naturally on Earth, so it would either have to be bred from neutron reactions (counteracting the potential advantage of aneutronic fusion), or mined from extraterrestrial sources.

The top several meters of the surface of the Moon is relatively rich in ^{3}He, on the order of 0.01 parts per million by weight, but mining this resource and returning it to Earth would be very difficult and expensive. ^{3}He could in principle be recovered from the atmospheres of the gas giant planets, Jupiter, Saturn, Neptune and Uranus, but this would be even more challenging.

The p –^{7}Li reaction has no advantage over p–^{11}B, given its somewhat lower cross section.

For the above reasons, most studies of aneutronic fusion concentrate on the reaction, p –^{11}B.

Despite the suggested advantages of aneutronic fusion, the vast majority of fusion research has gone toward D-T fusion because the technical challenges of hydrogen–boron (p –^{11}B) fusion are so formidable.

Hydrogen–boron fusion requires ion energies or temperatures almost ten times higher than those for D-T fusion. For any given densities of the reacting nuclei, the reaction rate for hydrogen-boron achieves its peak rate at around 600 keV (6.6 billion degrees Celsius or 6.6 gigakelvins).

In addition, the peak reaction rate of p–^{11}B is only one third that for D–T, requiring better plasma confinement.

Confinement is usually characterized by the time τ the energy must be retained so that the fusion power released exceeds the power required to heat the plasma.

Since the confinement properties of conventional fusion approaches, such as the tokamak and laser pellet fusion are marginal, most aneutronic proposals use radically different confinement concepts.

In every published fusion power plant design, the part of the plant that produces the fusion reactions is much more expensive than the part that converts the nuclear power to electricity. In that case, as indeed in most power systems, power density is a very important characteristic.

Doubling power density at least halves the cost of electricity. In addition, the confinement time required depends on the power density.

Lawrenceville Plasma Physics has published initial results and outlined a theory and experimental program for aneutronic fusion with the Dense Plasma Focus (DPF), building on earlier discussions. The private effort was initially funded by NASA's Jet Propulsion Laboratory. Support for other DPF aneutronic fusion investigations has come from the Air Force Research Laboratory.

Polywell fusion was pioneered by Robert W. Bussard and funded by the US Navy, uses inertial electrostatic confinement. Research continues at the company he founded, EMC2.

The Z-machine at Sandia National Laboratory, a z-pinch device, can produce ion energies of interest to hydrogen–boron reactions, up to 300 keV. Non-equilibrium plasmas usually have an electron temperature higher than their ion temperature, but the plasma in the Z machine has a special, reverted non-equilibrium state, where ion temperature is 100 times higher than electron temperature. These data represent a new research field, and indicate that Bremsstrahlung losses could be in fact lower than previously expected in such a design.

None of these efforts has yet tested its device with hydrogen–boron fuel, so the anticipated performance is based on extrapolating from theory, experimental results with other fuels and from simulations.

A picosond laser produced hydrogen–boron aneutronic fusions for a Russian team in 2005. However, the number of the resulting α particles (around 103 per laser pulse) was extremely low.

Aneutronic fusion reactions produce the overwhelming bulk of their energy in the form of charged particles instead of neutrons.

This means that energy could be converted directly into electricity by various techniques.

Many proposed direct conversion techniques are based on mature technology derived from other fields, such as microwave technology, and some involve equipment that is more compact and potentially cheaper than that involved in conventional thermal production of electricity.

In contrast, fusion fuels like deuterium-tritium (DT), which produce most of their energy in the form of neutrons, require a standard thermal cycle, in which the neutrons are used to boil water, and the resulting steam drives a large turbine and generator.

This equipment is sufficiently expensive that about 80% of the capital cost of a typical fossil-fuel electric power generating station is in the thermal conversion equipment.

Thus, fusion with DT fuels could not significantly reduce the capital costs of electric power generation even if the fusion reactor that produces the neutrons were cost-free. (Fuel costs would, however, be greatly reduced.) But according to proponents, aneutronic fusion with direct electric conversion could, in theory, produce electricity with reduced capital costs.

Direct conversion techniques can either be inductive, based on changes in magnetic fields, or electrostatic, based on making charged particles work against an electric field.

If the fusion reactor worked in a pulsed mode, inductive techniques could be used.

A sizable fraction of the energy released by aneutronic fusion would not remain in the charged fusion products but would instead be radiated as X-rays.

Some of this energy could also be converted directly to electricity.

Because of the photoelectric effect, X-rays passing through an array of conducting foils would transfer some of their energy to electrons, which can then be captured electrostatically.

Since X-rays can go through far greater thickness of material than electrons can, many hundreds or even thousands of layers would be needed to absorb most of the X-rays.

Dense plasma focus

A dense plasma focus (DPF) is a machine that produces, by electromagnetic acceleration and compression, a short-lived plasma that is so hot and dense that it can cause nuclear fusion and emit X-rays.

The electromagnetic compression of the plasma is called a pinch.

It was invented in the early 1960s by J.W. Mather and also independently by N.V. Filippov.

The plasma focus is similar to the high-intensity plasma gun device (HIPGD) (or just plasma gun), which ejects plasma in the form of a plasmoid, without pinching it.

Intense bursts of X-rays and charged particles are emitted, as are nuclear fusion neutrons, when operated using deuterium.

There is ongoing research that demonstrates potential applications as a soft X-ray source for next-generation microelectronics lithography, surface micromachining, pulsed X-ray and neutron source for medical and security inspection applications and materials modification, among others.

For nuclear weapons applications, dense plasma focus devices can be used as an external neutron source.

Other applications include simulation of nuclear explosions (for testing of the electronic equipment) and a short and intense neutron source useful for non-contact discovery or inspection of nuclear materials (uranium, plutonium).

An important characteristic of the dense plasma focus is that the energy density of the focused plasma is practically a constant over the whole range of machines, from sub-kilojoule machines to megajoule machines, when these machines are tuned for optimal operation.

This means that a small table-top-sized plasma focus machine produces essentially the same plasma characteristics (temperature and density) as the largest plasma focus. Of course the larger machine will produce the larger volume of focused plasma with a corresponding longer lifetime and more radiation yield.

Even the smallest plasma focus has essentially the same dynamic characteristics as larger machines, producing the same plasma characteristics and the same radiation products. This is due to the scalability of plasma phenomena.

See also plasmoid, the self-contained magnetic plasma ball that may be produced by a dense plasma focus.

The charged bank of electrical capacitors (also called a Marx bank or Marx generator) is switched onto the anode.

The gas breaks down. A rapidly rising electric current flows across the backwall electrical insulator, axisymmetrically, as depicted by the path (labeled 1).

The axisymmetric sheath of plasma current lifts off the insulator due to the interaction of the current with its own magnetic field (Lorentz force). The plasma sheath is accelerated axially, to position 2, and then to position 3, ending the axial phase of the device.

The whole process proceeds at many times the speed of sound in the ambient gas.

As the current sheath continues to move axially, the portion in contact with the anode slides across the face of the anode, axisymmetrically.

When the imploding front of the shock wave coalesces onto the axis, a reflected shock front emanates from the axis until it meets the driving current sheath which then forms the axisymmetric boundary of the pinched, or focused, hot plasma column.

A Filippov focus has a very short axial phase compared to a Mather focus.

A network of ten identical DPF machines operates in eight countries around the world.

This network produces research papers on topics including machine optimization & diagnostics (soft x-rays, neutrons, electron and ion beams), applications (microlithography, micromachining, materials modification and fabrication, imaging & medical, astrophysical simulation) as well as modeling & computation.

The network was organized by Sing Lee in 1986 and is coordinated by the Asian African Association for Plasma Training, AAAPT.

A simulation package, the Lee Model, has been developed for this network but is applicable to all plasma focus devices.

The code typically produces excellent agreement between computed and measured results, and is available for downloading as a Universal Plasma Focus Laboratory Facility.

The Institute for Plasma Focus Studies IPFS was founded on 25 February 2008 to promote correct and innovative use of the Lee Model code and to encourage the application of plasma focus numerical experiments. IPFS research has already extended numerically-derived neutron scaling laws to multi-megajoule experiments. These await verification.

Numerical experiments with the code have also resulted in the compilation of a global scaling law indicating that the well-known neutron saturation effect is better correlated to a scaling deterioration mechanism.

This is due to the increasing dominance of the axial phase dynamic resistance as capacitor bank impedance decreases with increasing bank energy (capacitance).

In principle, the resistive saturation could be overcome by operating the pulse power system at a higher voltage.

In Argentina there is an Inter-institutional Program for Plasma Focus Research since 1996, coordinated by a National Laboratory of Dense Magnetized Plasmas in Tandil, Buenos Aires.

The Program also cooperates with the Chilean Nuclear Energy Commission, and networks the Argentine National Energy Commission, the Scientific Council of Buenos Aires, the University of Center, the University of Mar del Plata, The University of Rosario, and the Institute of Plasma Physics of the University of Buenos Aires.

The program operates six Plasma Focus Devices, developing applications, in particular ultra-short tomography and substance detection by neutron pulsed interrogation.

Chile currently operates the facility SPEED-2, the largest Plasma Focus facility of the southern hemisphere. PLADEMA also contributed during the last decade with several mathematical models of Plasma Focus.

The thermodynamic model was able to develop for the first time design maps combining geometrical and operational parameters, showing that there is always an optimum gun length and charging pressure which maximize the neutron emission.

Currently there is a complete finite-elements code validated against numerous experiments, which can be used confidently as a design tool for Plasma Focus.

Since the beginning of 2009, a number of new plasma focus machines have been/are being commissioned including the INTI Plasma Focus in Malaysia, the NX3 in Singapore and the first plasma focus to be commissioned in a US university in recent times, the KSU Plasma Focus at Kansas State University which recorded its first fusion neutron emitting pinch on New Year's Eve 2009.

Several groups have proposed that fusion power based on the DPF could be viable, possibly even with low-neutron fuel cycles like p-B11.

The feasibility of net power from p-B11 in the DPF requires that the bremsstrahlung losses be reduced by quantum mechanical effects induced by the powerful magnetic field.

The high magnetic field will also result in a high rate of emission of cyclotron radiation, but at the densities envisioned, where the plasma frequency is larger than the cyclotron frequency, most of this power will be reabsorbed before being lost from the plasma.

Another advantage claimed is the capability of direct conversion of the energy of the fusion products into electricity, with an efficiency potentially above 70%. Experiments and computer simulations to investigate the capability of DPF for fusion power are underway at Lawrenceville Plasma Physics (LPP) under the direction of Eric Lerner, who explained his "Focus Fusion" approach in a 2007 Google Tech Talk.

Bibliography

[1] David Halliday, Robert, R., - *Physics, Part II,* Edit. John Wiley & Sons, Inc. - New York, London, Sydney, 1966;
[2] Petrescu F.I., *The movement of an electron around the atomic nucleus,* in ICOME 2010, Craiova, 2010.
[3] "Progress in Fusion". ITER. Retrieved 2010-02-15.
[4] "The National Ignition Facility: Ushering in a New Age for Science". National Ignition Facility. Retrieved 2009-09-13.
[5] "DOE looks again at inertial fusion as potential clean-energy source", David Kramer, Physics Today, March 2011, p 26
[6] The Most Tightly Bound Nuclei. Hyperphysics.phy-astr.gsu.edu. Retrieved on 2011-08-17.
[7] F. Winterberg "Conjectured Metastable Super-Explosives formed under High Pressure for Thermonuclear Ignition"
[8] Zhang, Fan; Murray, Stephen Burke; Higgins, Andrew (2005) "Super compressed detonation method and device to effect such detonation[dead link]"
[9] I.I. Glass and J.C. Poinssot "IMPLOSION DRIVEN SHOCK TUBE". NASA
[10] D.Sagie and I.I. Glass (1982) "Explosive-driven hemispherical implosions for generating fusion plasmas"
[11] T. Saito, A. K. Kudian and I. I. Glass "Temperature Measurements Of An Implosion Focus"
[12] S.E. Jones (1986). "Muon-Catalysed Fusion Revisited". Nature 321 (6066): 127–133. Bibcode 1986Natur.321..127J. DOI:10.1038/321127a0.
[13] Access: Desktop fusion is back on the table: Nature News. Nature.com. Retrieved on 2011-08-17.
[14] Supplementary methods for "Observation of nuclear fusion driven by a pyroelectric crystal". Main article Naranjo, B.; Gimzewski, J.K.; Putterman, S. (2005). "Observation of nuclear fusion driven by a pyroelectric crystal". Nature 434 (7037): 1115–1117. Bibcode 2005Natur.434.1115N. DOI:10.1038/nature03575. PMID 15858570.

CHAPTER II – SOME FEW SPECIFICATIONS ABOUT THE DOPPLER EFFECT TO THE ELECTROMAGNETIC WAVES

Introduction

The Doppler effect (or Doppler shift), named after Austrian physicist Christian Doppler who proposed it in 1842 in Prague, is the change in frequency of a wave for an observer moving relative to the source of the wave.

It is commonly heard when a vehicle sounding a siren or horn approaches, passes, and recedes from an observer.

The received frequency is higher (compared to the emitted frequency) during the approach, it is identical at the instant of passing by, and it is lower during the recession (See the Figure 1).

The relative changes in frequency can be explained as follows.

When the source of the waves is moving toward the observer, each successive wave crest is emitted from a position closer to the observer than the previous wave.

Therefore each wave takes slightly less time to reach the observer than the previous wave.

Therefore the time between the arrival of successive wave crests at the observer is reduced, causing an increase in the frequency.

While they are travelling, the distance between successive wave fronts is reduced; so the waves "bunch together".

Conversely, if the source of waves is moving away from the observer, each wave is emitted from a position farther from the observer than the previous wave, so the arrival time between successive waves is increased, reducing the frequency.

The distance between successive wave fronts is increased, so the waves "spread out".

For waves that propagate in a medium, such as sound waves, the velocity of the observer and of the source is relative to the medium in which the waves are transmitted.

The total Doppler Effect may therefore result from motion of the source, motion of the observer, or motion of the medium.

Each of these effects is analyzed separately.

For waves which do not require a medium, such as light or gravity in general relativity, only the relative difference in velocity between the observer and the source needs to be considered.

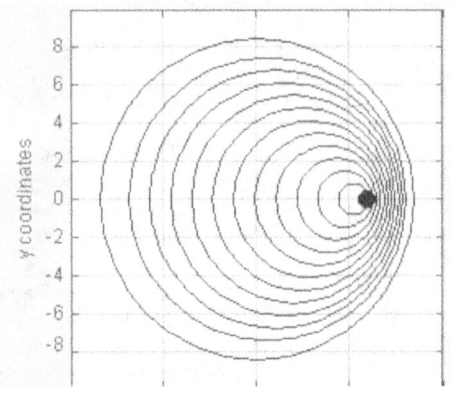

Fig. 1 *The Doppler Effect model*

Development

Doppler first proposed the effect in 1842 in his treatise "Über das farbige Licht der Doppelsterne und einiger anderer Gestirne des Himmels" (On the coloured light of the binary stars and some other stars of the heavens).

The hypothesis was tested for sound waves by Buys Ballot in 1845.

He confirmed that the sound's pitch was higher than the emitted frequency when the sound source approached him, and lower than the emitted frequency when the sound source receded from him.

Hippolyte Fizeau discovered independently the same phenomenon on electromagnetic waves in 1848 (in France, the effect is sometimes called "l'Effet Doppler-Fizeau" but that name was not adopted by the rest of the world as Fizeau's discovery was three years after Doppler's).

In Britain, John Scott Russell made an experimental study of the Doppler Effect (1848).

Craig Bohren pointed out in 1991 that some physics textbooks erroneously state that the observed frequency increases as the object approaches an observer and then decreases only as the object passes the observer.

In most cases, the observed frequency of an approaching object declines monotonically from a value above the emitted frequency, through a value equal to the emitted frequency when the object is closest to the observer, and to values increasingly below the emitted frequency as the object recedes from the observer.

Bohren proposed that this common misconception might occur because the intensity of the sound increases as an object approaches an observer and decreases once it passes and recedes from the observer and that this change in intensity is misperceived as a change in frequency.

Higher sound pressure levels make for a small decrease in perceived pitch in low frequency sounds, and for a small increase in perceived pitch for high frequency sounds.

Application

Sirens

The siren on a passing emergency vehicle will start out higher than its stationary pitch, slide down as it passes, and continue lower than its stationary pitch as it recedes from the observer. Astronomer John Dobson explained the effect thus: "The reason the siren slides is because it doesn't hit you."

In other words, if the siren approached the observer directly, the pitch would remain constant (as $v_{s,r}$ is only the radial component) until the vehicle hit him, and then immediately jump to a new lower pitch.

Because the vehicle passes by the observer, the radial velocity does not remain constant, but instead varies as a function of the angle between his line of sight and the siren's velocity:

$$v_r = v_s \cdot \cos\theta$$

where v_s is the velocity of the object (source of waves) with respect to the medium, and θ is the angle between the object's forward velocity and the line of sight from the object to the observer.

Astronomy

Redshift of spectral lines in the optical spectrum of a supercluster of distant galaxies (right), as compared to that of the Sun (left)

The Doppler Effect for electromagnetic waves such as light is of great use in astronomy and results in either a so-called red shift or blue shift. It has been used to measure the speed at which stars and galaxies are approaching or receding from us, that is, the radial velocity.

This is used to detect if an apparently single star is, in reality, a close binary and even to measure the rotational speed of stars and galaxies.

The use of the Doppler Effect for light in astronomy depends on our knowledge that the spectra of stars are not continuous.

They exhibit absorption lines at well defined frequencies that are correlated with the energies required to excite electrons in various elements from one level to another.

The Doppler Effect is recognizable in the fact that the absorption lines are not always at the frequencies that are obtained from the spectrum of a stationary light source.

Since blue light has a higher frequency than red light, the spectral lines of an approaching astronomical light source exhibit a blue shift and those of a receding astronomical light source exhibit a redshift.

Among the nearby stars, the largest radial velocities with respect to the Sun are +308 km/s (BD-15°4041, also known as LHS 52, 81.7 light-years away) and -260 km/s (Woolley 9722, also known as Wolf 1106 and LHS 64, 78.2 light-years away).

Positive radial velocity means the star is receding from the Sun, negative that it is approaching.

Temperature measurement

Another use of the Doppler Effect, which is found mostly in plasma physics and astronomy, is the estimation of the temperature of a gas (or ion temperature in a plasma) which is emitting a spectral line.

Due to the thermal motion of the emitters, the light emitted by each particle can be slightly red or blue-shifted, and the net effect is a broadening of the line.

This line shape is called a Doppler profile and the width of the line is proportional to the square root of the temperature of the emitting species, allowing a spectral line (with the width dominated by the Doppler broadening) to be used to infer the temperature.

Radar

The Doppler Effect is used in some types of radar, to measure the velocity of detected objects.

A radar beam is fired at a moving target — e.g. a motor car, as police use radar to detect speeding motorists — as it approaches or recedes from the radar source.

Each successive radar wave has to travel farther to reach the car, before being reflected and re-detected near the source.

As each wave has to move farther, the gap between each wave increases, increasing the wavelength.

In some situations, the radar beam is fired at the moving car as it approaches, in which case each successive wave travels a lesser distance, decreasing the wavelength.

In either situation, calculations from the Doppler Effect accurately determine the car's velocity.

Moreover, the proximity fuze, developed during World War II, relies upon Doppler radar to explode at the correct time, height, distance, etc.

Medical imaging and blood flow measurement

Color flow ultrasonography (Doppler) of a carotid artery - scanner and screen

An echocardiogram can, within certain limits, produce accurate assessment of the direction of blood flow and the velocity of blood and cardiac tissue at any arbitrary point using the Doppler Effect.

One of the limitations is that the ultrasound beam should be as parallel to the blood flow as possible.

Velocity measurements allow assessment of cardiac valve areas and function, any abnormal communications between the left and right side of the heart, any leaking of blood through the valves (valvular regurgitation), and calculation of the cardiac output.

Contrast-enhanced ultrasound using gas-filled microbubble contrast media can be used to improve velocity or other flow-related medical measurements.

Although "Doppler" has become synonymous with "velocity measurement" in medical imaging, in many cases it is not the frequency shift (Doppler shift) of the received signal that is measured, but the phase shift (when the received signal arrives).

Velocity measurements of blood flow are also used in other fields of medical ultrasonography, such as obstetric ultrasonography and neurology.

Velocity measurement of blood flow in arteries and veins based on Doppler Effect is an effective tool for diagnosis of vascular problems like stenosis.

Flow measurement

Instruments such as the laser Doppler velocimeter (LDV), and acoustic Doppler velocimeter (ADV) have been developed to measure velocities in a fluid flows.

The LDV emits a light beam and the ADV emits an ultrasonic acoustic burst, and measure the Doppler shift in wavelengths of reflections from particles moving with the flow.

The actual flow is computed as a function of the water velocity and phase.

This technique allows non-intrusive flow measurements, at high precision and high frequency.

Velocity profile measurement

Developed originally for velocity measurements in medical applications (blood flow), Ultrasonic Doppler Velocimetry (UDV) can measure in real time complete velocity profile in almost any liquids containing particles in suspension such as dust, gas bubbles, emulsions.

Flows can be pulsating, oscillating, laminar or turbulent, stationary or transient.

This technique is fully non-invasive.

Satellite communication

Fast moving satellites can have a Doppler shift of dozens of kilohertz relative to a ground station.

The speed, thus magnitude of Doppler Effect, changes due to earth curvature.

Dynamic Doppler compensation, where the frequency of a signal is changed multiple times during transmission, is used so the satellite receives a constant frequency signal.

Underwater acoustics

In military applications the Doppler shift of a target is used to ascertain the speed of a submarine using both passive and active sonar systems.

As a submarine passes by a passive sonobuoy, the stable frequencies undergo a Doppler shift, and the speed and range from the sonobuoy can be calculated.

If the sonar system is mounted on a moving ship or another submarine, then the relative velocity can be calculated.

A sonobuoy (a portmanteau of sonar and buoy; see the Figure 2) is a relatively small (typically 5 inches / 13 centimeters, in diameter and 3 ft/91 cm long) expendable sonar system that is dropped/ejected from aircraft or ships conducting anti-submarine warfare or underwater acoustic research.

The buoys are ejected from aircraft in canisters and deploy upon water impact.

An inflatable surface float with a radio transmitter remains on the surface for communication with the aircraft, while one or more hydrophone sensors and stabilizing equipment descend below the surface to a selected depth that is variable, depending on environmental conditions and the search pattern.

The buoy relays acoustic information from its hydrophone(s) via UHF/VHF radio to operators onboard the aircraft.

Fig. 2 *Sonobuoy being loaded onto an USN P-3C Orion aircraft*

Vibration measurement

A laser Doppler vibrometer (LDV) is a non-contact method for measuring vibration.

The laser beam from the LDV is directed at the surface of interest, and the vibration amplitude and frequency are extracted from the Doppler shift of the laser beam frequency due to the motion of the surface.

A laser Doppler vibrometer (LDV) is a scientific instrument that is used to make non-contact vibration measurements of a surface.

The laser beam from the LDV is directed at the surface of interest, and the vibration amplitude and frequency are extracted from the Doppler shift of the laser beam frequency due to the motion of the surface.

The output of an LDV is generally a continuous analog voltage that is directly proportional to the target velocity component along the direction of the laser beam.

Some advantages of an LDV over similar measurement devices such as an accelerometer are that the LDV can be directed at targets that are difficult to access, or that may be too small or too hot to attach a physical transducer.

Also, the LDV makes the vibration measurement without mass-loading the target, which is especially important for MEMS devices.

The relativistic Doppler Effect

The relativistic Doppler Effect (Figure 3) is the change in frequency (and wavelength) of light, caused by the relative motion of the source and the observer (as in the classical Doppler Effect), when taking into account effects described by the special theory of relativity.

The relativistic Doppler Effect is different from the non-relativistic Doppler Effect as the equations include the time dilation effect of special relativity and do not involve the medium of propagation as a reference point.

They describe the total difference in observed frequencies and possess the required Lorentz symmetry.

Fig. 3 *A source of light waves moving to the right with velocity 0.7c. The frequency is higher on the right, and lower on the left.*

The photoacoustic Doppler Effect

The photoacoustic Doppler Effect, as its name implies, is one specific kind of Doppler Effect, which occurs when an intensity modulated light wave induces a photoacoustic wave on moving particles with a specific frequency.

The observed frequency shift is a good indicator of the velocity of the illuminated moving particles. A potential biomedical application is measuring blood flow.

Specifically, when an intensity modulated light wave is exerted on a localized medium, the resulting heat can induce an alternating and localized pressure change.

This periodic pressure change generates an acoustic wave with a specific frequency.

Among various factors that determine this frequency, the velocity of the heated area and thus the moving particles in this area can induce a frequency shift proportional to the relative motion.

Thus, from the perspective of an observer, the observed frequency shift can be used to derive the velocity of illuminated moving particles.

SOME SPECIFICATIONS ABOUT THE MODERN DOPPLER EFFECT

Doppler Effect Found Even at Molecular Level -- 169 Years After Its Discovery

Science Daily (May 10, 2011) — Whether they know it or not, anyone who's ever gotten a speeding ticket after zooming by a radar gun has experienced the Doppler Effect -- a measurable shift in the frequency of radiation based on the motion of an object, which in this case is your car doing 45 miles an hour in a 30-mph zone.

But for the first time, scientists have experimentally shown a different version of the Doppler Effect at a much, much smaller level -- the rotation of an individual molecule. Prior to this such an effect had been theorized, but it took a complex experiment with a synchrotron to prove it's for real.

"Some of us thought of this some time ago, but it's very difficult to show experimentally," said T. Darrah Thomas, a professor emeritus of chemistry at Oregon State University and part of an international research team that just announced its findings in Physical Review Letters, a professional journal.

Most illustrations of the Doppler Effect are called "translational," meaning the change in frequency of light or sound when one object moves away from the other in a straight line, like a car passing a radar gun. The basic concept has been understood since an Austrian physicist named Christian Doppler first proposed it in 1842. But a similar effect can be observed when something rotates as well, scientists say. "There is plenty of evidence of the rotational Doppler Effect in large bodies, such as a spinning planet or galaxy," Thomas said. "When a planet rotates, the light coming from it shifts to higher frequency on the side spinning toward you and a lower frequency on the side spinning away from you. But this same basic force is at work even on the molecular level."

In astrophysics, this rotational Doppler Effect has been used to determine the rotational velocity of things such as planets. But in the new study, scientists from Japan, Sweden, France and the United States provided the first experimental proof that the same thing happens even with molecules. At this tiny level, they found, the rotational Doppler Effect can be even more important than the linear motion of the molecules, the study showed.

The findings are expected to have application in a better understanding of molecular spectroscopy, in which the radiation emitted from molecules is used to study their makeup and chemical properties. It is also relevant to the study of high energy electrons, Thomas said. "There are some studies where a better understanding of this rotational Doppler Effect will be important," Thomas said. "Mostly it's just interesting. We've known about the Doppler Effect for a very long time but until now have never been able to see the rotational Doppler Effect in molecules."

References

[1] T. D. Thomas, E. Kukk, K. Ueda, T. Ouchi, K. Sakai, T. X. Carroll, C. Nicolas, O. Travnikova, and C. Miron. **Experimental observation of rotational Doppler broadening in a molecular system**. *Physical Review Letters*, Accepted Apr 12, 2011.

[2] Oregon State University (2011, May 10). Doppler Effect found even at molecular level -- 169 years after its discovery. *Science Daily*. http://www.sciencedaily.com/releases/2011/05/110510134112.htm

Did Scientists Break the Speed of Light?

CERN Records Sub-Atomic Particle Speeds

September 23, 2011 - Albert Einstein is responsible for many of the longest-standing laws of physics, including the famous theory of special relativity. Nearly all of modern physics and astronomy is based upon the concept that nothing can travel faster than the speed of light. It appears that Einstein may have been proven wrong thanks to a new study completed by CERN (the European Organization for Nuclear Research).

CERN claims that they have recorded sub-atomic particles traveling at a speed faster than light. The OPERA experiment consisted of firing 15,000 beams of neutrinos from Geneva, Switzerland to Gran Sasso, Italy over a period of 3 years. The sensors in Italy registered that the particles reached the target 60 nanoseconds faster than the speed of light. According to modern knowledge, this is an impossible feat. CERN scientists asked for others to confirm their research by reproducing the results, and Fermilab in Chicago is already attempting to recreate the experiment.

Despite asking for a double check, the CERN scientists seem very sure of this finding. "We have high confidence in our results, stated spokesman Antonio Ereditato.

"We have checked and rechecked for anything that could have distorted our measurements but we found nothing." If these findings are indeed accurate, they will shake the very foundation of Einstein's theory.

While some people might think that 60 nanoseconds is insignificant, that little amount may have serious implications in the possibilities of light speed travel or even time travel. However, the neutrino is still quite mysterious to scientists, so further research is necessary to prove exactly what the OPERA experiment proved.

It's pretty amazing how little we actually know about our universe. Something new is often discovered that completely changes our perceptions of science and reality. I'm really looking forward to seeing what the scientific community discovers if the light speed barrier is actually broken.

References

[1] http://www.chacha.com/topic/light-speed/gallery/995/did-scientists-break-the-speed-of-light

Meteorologists Invent Better Way to Monitor Hurricane Strength

September 1, 2007 — Meteorologists have developed a new method for analyzing hurricane strength. A series of mathematical formulas transform data from Doppler radars into a 3-D picture of storm intensity every 6 minutes. Because of the rapid updates, the technique increases meteorologists' ability to capture sudden, dangerous changes in hurricanes.

The strongest hurricane to hit the U.S. in more than a decade -- killing ten people -- causing thirteen-billion dollars in damage. Its arrival was expected. Its intensity: an absolute surprise.

We are in the middle of hurricane season again and meteorologists are rushing to test a new way to track a storm's intensity. Scientists now know, as hurricane Charlie approached Florida three years ago, Floridians were preparing for the storm with obsolete information.

Charlie landed with 25-percent more intensity than predicted. It's a scenario that could forever be avoided with a new tracking system.

'[Hurricane Charlie] rapidly intensified from category two to category four in roughly three hours,' said Wen-Chau Lee, NCAR meteorologist.

If only meteorologists knew then what they know now. Now, meteorologists at the national center for atmospheric research have a new software tool called "VORTRAC." It slices through approaching hurricanes to reveal a three-dimensional view of the storm and just how intense it will be. The result looks a lot like the layers of a sliced onion.

'When you cut an onion in half you see different rings. Basically what we do, we dissect a hurricane into different rings,' Lee said.

VORTRAC combines wind measurements from the Doppler radar closest to the eye of the storm with existing hurricane data to show a 3-D view of the wind. Lee said he looks forward to putting his tracking system to the test in the U.S. when the next hurricane heads our way.

Because of the limited range of Doppler radars, VORTRAC works only for hurricanes within about 120 miles of land. But that could help monitor the critical 10 to 15 hours before landfall. The National Hurricane Center is testing the system currently and expects it to be ready for use in about two years.

BACKGROUND: Forecasters are testing a new technique called VORTRAC -- Vortex Objective Radar Tracking and Circulation -- that provides a detailed 3D view of an approaching hurricane every six minutes and allows them to determine whether the storm is gathering strength as it nears land. Then they can quickly alert coastal communities if it suddenly strengthens.

HOW IT WORKS: Developed by researchers at the National Center for Atmospheric Research (NCAR), the technique relies on the existing network of Doppler radars along the southeast coast to closely monitor hurricanes winds. Any radar can measure winds blowing toward or away from it, but no single radar could provide a 3D picture of hurricane winds until now.

The NCAR scientists developed a series of mathematical formulas that combine data from a single radar near the center of a landfalling storm with general knowledge of Atlantic hurricane structure in order to map the approaching system's winds in three dimensions. The technique also infers the barometric pressure in the eye of the hurricane, a very reliable index of its strength. However, because of the limited range of Doppler radars, VORTRAC works only for hurricanes that are within about 120 miles of land. In the future, it might be possible to use VORTRAC to help improve long-range hurricane forecasts by using data from airborne radars to glean detailed information about a hurricane that is far out to sea.

ABOUT HURRICANES: A hurricane is a type of tropical cyclone, a low-pressure system that usually forms in the tropics and has winds that circulate counterclockwise near the earth's surface. Storms are considered hurricanes when their wind speeds surpass 74 MPH. Every hurricane arises from the combination of warm water and moist warm air. Tropical thunderstorms drift out over warm ocean waters and encounter winds coming in from near the equator. Warm, moist air from the ocean surface rises rapidly, encounters cooler air, and condensed into water vapor to form storm clouds, releasing heat in the process.

This heat causes the condensation process to continue, so that more and more warm moist air is drawn into the developing storm, creating a wind pattern that spirals around the relatively calm center, or eye, of the storm, much like water swirling down a drain. The winds keep circling and accelerating to form a classic cyclone pattern.

WHAT IS DOPPLER RADAR: Doppler radar uses a well-known effect of light called the Doppler shift. Just as a train whistle will sound higher as it approaches a platform and then become lower in pitch as it moves away, light emitted by a moving object is perceived to increase in frequency (a blue shift) if it is moving toward the observer; if the object is moving away from us, it will be shifted toward the red end of the spectrum. Doppler radar sends out radio waves that bounce off objects in the air, such as raindrops or snow crystals, and then measures how much the frequency changes in returning radio waves to better determine wind direction and speed.

References

[1] 3D Hurricane Tracking. *Science Daily*.

http://www.sciencedaily.com/videos/2007/0901-3d_hurricane_tracking.htm

3D Doppler Ultrasound Helps Identify Breast Cancer

ScienceDaily (Oct. 21, 2008) — Three-dimensional (3-D) power Doppler ultrasound helps radiologists distinguish between malignant and benign breast masses, according to a new study being published in the November issue of Radiology.

"Using 3-D scans promises greater accuracy due to more consistent sampling over the entire tumor," said lead author, Gerald L. LeCarpentier, Ph.D., assistant professor in the Department of Radiology at University of Michigan in Ann Arbor. "Our study shows that 3-D power Doppler ultrasound may be useful in the evaluation of some breast masses."

Malignant breast masses often exhibit increased blood flow compared to normal tissue or benign masses. Using 3-D power Doppler ultrasound, radiologists are able to detect vessels with higher flow speeds, which likely indicate cancer.

For the study, Dr. Le Carpentier and colleagues studied 78 women between the ages of 26 and 70 who were scheduled for biopsy of a suspicious breast mass. Each of the women underwent a 3-D Doppler ultrasound exam followed by core or excisional biopsy of the breast.

The results showed that 3-D power Doppler ultrasound was highly accurate in identifying malignant breast tumors. When combined with age-based assessment and gray scale visual analysis, 3-D Doppler showed a sensitivity of 100 percent in identifying cancerous tumors and a specificity of 86 percent in excluding benign tumors. "Using speed-weighted 3-D power Doppler ultrasound, higher flow velocities in the malignant tumor-feeding vessels may be detected, whereas vessels with slower flow velocities in surrounding benign masses may be excluded," Dr. Le Carpentier said.

References

[1] 3-D Doppler Ultrasound Helps Identify Breast Cancer. *Science Daily*. http://www.sciencedaily.com/releases/2008/10/081021093933.htm

[2] Le Carpentier et al. Suspicious Breast Lesions: Assessment of 3D Doppler US Indexes for Classification in a Test Population and Fourfold Cross-Validation Scheme. Radiology, 2008; 249 (2): 463 DOI: 10.1148/radiol.2492060888

Perfusion In Burn Injuries Rapidly Determined By Using Improved Laser-Doppler Technology, Hospital Test Shows

ScienceDaily (Dec. 16, 2007) — The perfusion of a burn injury can now rapidly be determined by using a new technique developed by scientists of the University of Twente. Using the perfusion image made by a laser and an ultra fast camera, doctors will be able to determine whether a burn needs surgery. The new measuring device, developed under supervision of Dr. Wiendelt Steenbergen of the Biophysical Engineering group, has been successfully tested at the hospital Martini Ziekenhuis in Groningen.

Tests in hospital show that the system is perfectly capable of measuring differences in perfusion in burn wounds; patients and medical staff are positive about the high speed of the system and the quality of the images.

A burn that shows good perfusions, has a better chance of healing by itself: no surgery is needed. In many cases, the visual inspection is not sufficient to take a decision on the necessity of surgery. This can lead to unnecessary surgery or, on the other hand, to unwanted delays when surgery is the best option.

Compared to current perfusion measurements, the new technique is much faster. Scanning techniques take minutes of time for some square centimeters of skin, during which time the patient is not allowed to move. The new technique will be capable of imaging an entire surface of ten by ten centimeter in a fraction of a second.

Doppler Effect

In order to reach this high speed, the entire surface is lit at once using a wide laser beam. A high speed camera, capable of taking 27000 shots per second, takes images of the tissue. Whenever laser light is scattered by moving rood blood cells, this is visible in the intensity of the pixels; due to the Doppler effect, a color shift will be visible. From the resulting 'movie' of the tissue, a perfusion image can be made.

Apart from this promising application in determining perfusion in burn injuries, Wiendelt Steenbergen predicts other applications, for example in evaluating the uptake of medication through the skin, or in testing allergic reactions. In evaluating diabetic micro circulation problems, the new technique could be an attractively fast alternative to current approaches as well.

Patent

Current market leader in Laser-Doppler equipment for perfusion imaging, Perimed AB from Sweden, has shown interest in the new technique. The Swedish company has signed a contract with Dutch Technology Foundation STW, for acquiring the patent on the technique. STW finances Steenbergen's work.

The research has been done within the Biophysical Engineering Group of the University of Twente, which is part of the BMTI Institute for Biomedical Technology.

References

[1] Perfusion In Burn Injuries Rapidly Determined By Using Improved Laser-Doppler Technology, Hospital Test Shows. *Science Daily*.

http://www.sciencedaily.com/releases/2007/12/071216130313.htm

[2] University of Twente (2007, December 16). Perfusion In Burn Injuries Rapidly Determined By Using Improved Laser-Doppler Technology, Hospital Test Shows. *ScienceDaily*. Retrieved January 20, 2012, from http://www.sciencedaily.com/releases/2007/12/071216130313.htm

Doppler Radars Help Increase Monsoon Rainfall Prediction Accuracy

ScienceDaily (Oct. 5, 2010) — Doppler weather radar will significantly improve forecasting models used to track monsoon systems influencing the monsoon in and around India, according to a research collaboration including Purdue University, the National Center for Atmospheric Research and the Indian Institute of Technology Delhi. Dev Niyogi, a Purdue associate professor of agronomy and earth and atmospheric sciences, said modeling of a monsoon depression track can have a margin of error of about 200 kilometers for landfall, which can be significant for storms that produce as much as 20-25 inches of rain as well as inland floods and fatalities.

"When you run a forecast model, how you represent the initial state of the atmosphere is critical. Even if Doppler radar information may seem highly localized, we find that it enhances the regional atmospheric conditions, which, in turn, can significantly improve the dynamic prediction of how the monsoon depression will move as the storm makes landfall," Niyogi said. "It certainly looks like a wise investment made in Doppler radars can help in monsoon forecasting, particularly the heavy rain from monsoon processes."

Niyogi, U.C. Mohanty, a professor in the Centre for Atmospheric Sciences at the Indian Institute of Technology, and Mohanty's doctoral student, Ashish Routray, collaborated with scientists at the National Center for Atmospheric Research and gathered information such as radial velocity and reflectivity from six Doppler weather radars that were in place during storms. Using the Weather Research and Forecasting Model, they found that incorporating the Doppler radar-based information decreased the error of the monsoon depression's landfall path from 200 kilometers to 75 kilometers.

Monsoons account for 80 percent of the rain India receives each year. Mohanty said more accurate predictions could better prepare people for heavy rains that account for a number of deaths in a monsoon season.

"Once a monsoon depression passes through, it can cause catastrophic floods in the coastal areas of India," Mohanty said. "Doppler radar is a very useful tool to help assess these things."

The researchers modeled monsoon depressions and published their findings in the Quarterly Journal of the Royal Meteorological Society. Future studies will incorporate more simulations and more advanced models to test the ability of Doppler radar to track monsoon processes. Niyogi said the techniques and tools being developed also could help predict landfall of tropical storm systems that affect the Caribbean and the United States. The National Science Foundation CAREER program, U.S. Agency for International Development and the Ministry of Earth Sciences in India funded the study.

References

[1] Doppler Radars Help Increase Monsoon Rainfall Prediction Accuracy. *Science Daily*.

http://www.sciencedaily.com/releases/2010/10/101005171044.htm

[2] Purdue University (2010, October 5). Doppler radars help increase monsoon rainfall prediction accuracy. *ScienceDaily*. Retrieved January 20, 2012, from http://www.sciencedaily.com/releases/2010/10/101005171044.htm

Explain: the Doppler Effect

The same phenomenon behind changes in the pitch of a moving ambulance's siren is helping astronomers locate and study distant planets.

Many students learn about the Doppler effect in physics class, typically as part of a discussion of why the pitch of a siren is higher as an ambulance approaches and then lower as the ambulance passes by. The effect is useful in a variety of different scientific disciplines, including planetary science: Astronomers rely on the Doppler Effect to detect planets outside of our solar system, or exoplanets. To date, 442 of the 473 known exoplanets have been detected using the Doppler Effect, which also helps planetary scientists glean details about the newly found planets.

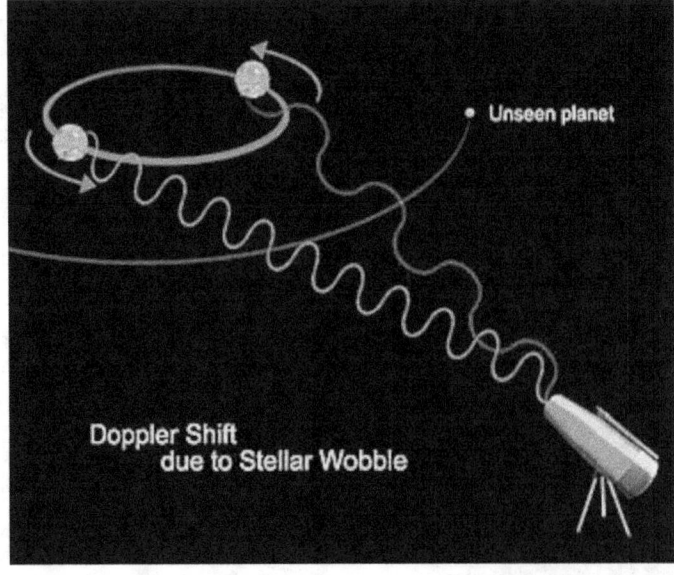

Fig. 4

The Doppler Effect, or Doppler shift, describes the changes in frequency of any kind of sound or light wave produced by a moving source with respect to an observer. Waves emitted by an object traveling toward an observer get compressed — prompting a higher frequency — as the source approaches the observer. In contrast, waves emitted by a source traveling away from an observer get stretched out.

In astronomy, that source can be a star that emits electromagnetic waves; from our vantage point, Doppler shifts occur as the star orbits around its own center of mass and moves toward or away from Earth. These wavelength shifts can be seen in the form of subtle changes in its spectrum, the rainbow of colors emitted in light. When a star moves toward us, its wavelengths get compressed, and its spectrum becomes slightly bluer.

When the star moves away from us, its spectrum looks slightly reddened. To observe the so-called red shifts and blue shifts over time, planetary scientists use a high-resolution prism-like instrument known as a spectrograph that separates incoming light waves into different colors. In every star's outer layer, there are atoms that absorb light at specific wavelengths, and this absorption appears as dark lines in the different colors of the star's spectrum that are recorded from the light emanating from the star. Researchers use the shifts in these lines as convenient markers by which to measure the size of the Doppler shift.

If the star exists by itself — that is, if there is no exoplanet or companion star in its stellar system — then there will be no change in the pattern of its Doppler shifts over time. But if there is a planet or companion star in the system, the gravitational pull of this unseen body or star will perturb the host star's movement at certain parts of its orbit, producing a noticeable change in the overall pattern and size of Doppler shifts over time. In other words, the pattern of a star's Doppler shifts can change over time as a result of gravity affecting the star's motion. "If this shift is large, then it must be caused by another star pulling it, but if this shift is small, then it is likely caused by a low-mass body like an exoplanet," explains Joshua Winn, an assistant professor in MIT's Department of Physics. As part of his work at MIT's Kavli Institute for Astrophysics and Space Research, Winn studies the relationship between an exoplanet's orbit and its parent star's rotation for clues about how the planet may have formed.

How a planet's Doppler shift changes over time can also shed light on the planet's orbital period (the length of its "year"), the shape of its orbit and its minimum possible mass. Recently, Kavli postdoc Simon Albrecht used the Doppler Effect to detect color shifts in the light absorbed by an exoplanet, which indicated strong winds in the planet's atmosphere.

Doppler shifts are used in many fields besides astronomy. By sending radar beams into the atmosphere and studying the changes in the wavelengths of the beams that come back, meteorologists use the Doppler Effect to detect water in the atmosphere. The Doppler phenomenon is also used in healthcare with echocardiograms that send ultrasound beams through a body to measure changes in blood flow to make sure that a heart valve is working properly or to diagnose vascular diseases. Police also rely on the Doppler Effect when they use a radar gun to bounce radio beams off of your car; the change in frequency between the directed and reflected beams provides a measure of your car's speed.

References

[1] Explain: the Doppler effect. MITnews (Massachusetts Institute of Technology).

http://web.mit.edu/newsoffice/2010/explained-doppler-0803.html

SOME FEW SPECIFICATIONS ABOUT THE DOPPLER EFFECT TO THE ELECTROMAGNETIC WAVES

The Doppler Effect [1-3] represents the frequency variation of the waves, received by an observer which is drawing (coming), respectively it's removing (going), from a wave spring (source).

If a bright spring is drawing to an observer, the frequency of waves received by the observer is bigger than the emitted frequency of source, such that the respective spectral lines are moving to violet.

On the contrary, if the light source is removing from the observer, the spectral lines are moving to red.

One proposes to study the Doppler Effect for the light waves, generally for the electromagnetic waves.

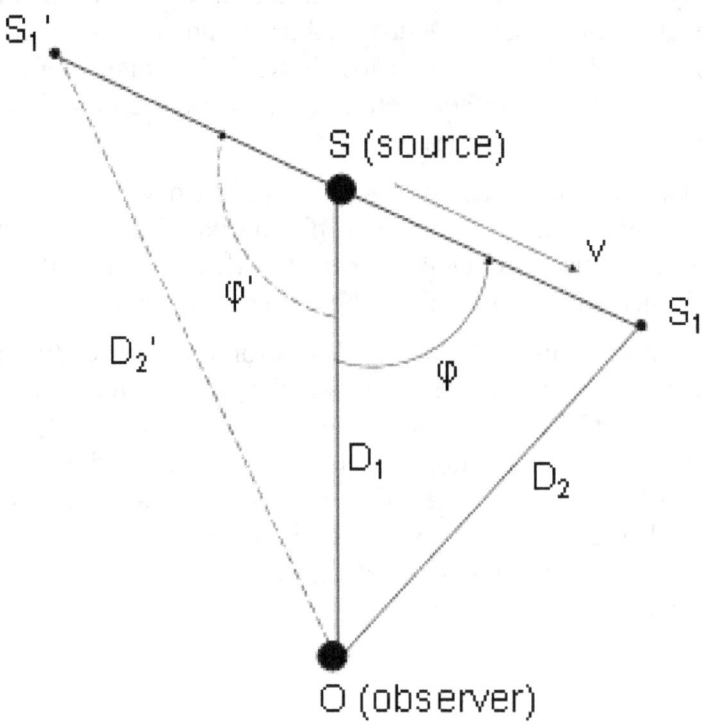

Fig. 5. *The waves received by an observer O from a waves source S, which is moving in relation with the observer, by the direction SS_1*

The new relations

We wish to calculate the period (T [s]) of the waves received by an observer O (figure 1) from a waves source S, which is moving in relation with the observer, on the direction SS_1 with the relative speed v [m/s] [1, 2].

T_0 [s] is the period of waves emitted by the source S.

At the moment t_0 [s], determinate by the observer O, from the source S bend a bright wave; this wave traverse the distance $D_1=SO$ [m] and arrive in O at the moment t_1 [s].

$$t_1 = t_0 + \frac{D_1}{c} \qquad (1)$$

where c is the light speed in vacuum: $c \cong 3.10^8$ [m/s].

After a T_0 period, from the source S (arrived now in S_1), from the source S_1 starts a second wave. The distance SS_1 [m] is:

$$SS_1 = v.T_0 \qquad (2)$$

The observer O, receive the second waves at the moment t_2 [s]:

$$t_2 = t_0 + T_0 + \frac{D_2}{c} \qquad (3)$$

The period T is equal with the difference between the two moments.

$$T = t_2 - t_1 = T_0 + \frac{D_2 - D_1}{c} \qquad (4)$$

The angle φ [rad] between the two vectors, SS_1 and SO is known and the distance $D_1=SO$ is known as well. With the COS theorem in the certain triangle SOS_1 one obtains the distance D_2 [m]:

$$D_2 = \sqrt{D_1^2 + SS_1^2 - 2.D_1.SS_1.\cos\varphi} \qquad (5)$$

With SS_1 from (2) the relation (5), become the expression (6).

$$D_2 = \sqrt{D_1^2 + v^2 T_0^2 - 2 D_1 v T_0 \cos \varphi} \qquad (6)$$

With the expression (6) in relation (4) one obtains the form (7).

$$T = T_0 + \frac{\sqrt{D_1^2 + v^2 T_0^2 - 2 D_1 v T_0 \cos \varphi} - D_1}{c} \qquad (7)$$

The relation (7) can be put in the form (8).

$$T = T_0 \left(1 + \beta \frac{v T_0 - 2 D_1 \cos \varphi}{\sqrt{D_1^2 + v^2 T_0^2 - 2 D_1 v T_0 \cos \varphi} + D_1}\right) \qquad (8)$$

where β is the ratio between the two speed, v and c:

$$\beta = \frac{v}{c} \qquad (9)$$

Presents the classical relation (10)

The classical relation (10) is very simply, but it's an approximate relation [2-3].

The expression (8) is more difficult but it's a very exact relation. It can be put in the forms (18), (19) and finally (20).

$$\frac{\gamma_0}{\gamma} = \frac{T}{T_0} = 1 - \beta \cos \varphi \qquad (10)$$

Some aspects

a) When the source S is removing from the observer, the angle φ (see the figure 1) take the values (φ') comprised between 90^0 and 180^0, cosφ become negative, the numerator of expression (8) become positive and the period of observer O (T) it'll be always bigger than T_0 (the period of source): $T > T_0$ and $\gamma < \gamma_0$ (the spectral lines are red).

When the source S is drawing to the observer, the angle $\varphi \in [0^0, 90^0)$ and cosφ>0. In this case one analyzes (11) the numerator of expression (8) and one can have two case (b and c) [1]:

$$N = v.T_0 - 2.D_1.\cos\varphi \tag{11}$$

b) If N<0, then $v.T_0 < 2.D_1.\cos\varphi$ or

$$\cos\varphi > \frac{v.T_0}{2.D_1} \tag{12}$$

and T<T₀, or y>y₀ (the spectral lines are violet) [1].

c) If N>0, then

$$\cos\varphi < \frac{v.T_0}{2.D_1} \tag{13}$$

and T>T₀, or y<y₀ (the spectral lines are red).

This case it wasn't known by the classical expression (10) [1].

d) The most interesting case is then when the angle $\varphi=90^0$, and $\cos\varphi=0$, when the source is moving perpendicular at the axle SO (see the figure 6). In this case the relation (8), become the expression (14).

$$T = T_0\left(1 + \frac{\beta.v.T_0}{\sqrt{D_1^2 + v^2.T_0^2} + D_1}\right) \tag{14}$$

T>T₀ and y<y₀ (the spectral lines are red) [1].

This fact can't be seen by the classical relation (10) which (for the $\varphi=90^0$), takes the form (15):

$$T = T_0 \tag{15}$$

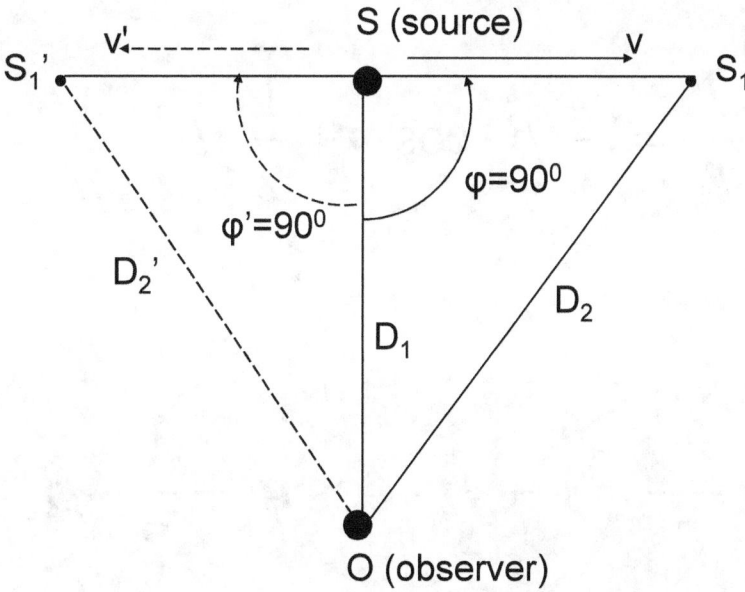

Fig. 6. *The waves received by an observer O from a waves source S when the source is moving perpendicular at the axle SO (it is a particular situation)*

The classical approximate relation (10, form 15) can't foresee the Doppler Effect for this case, but the effect virtually exist, and for this reason it was introduced the relativity effect (or the Lorentz transformation), where the period T_0 takes the form T_0/α (see [1]), and the relation (15) takes the form (16) [2, 3]:

$$T = \frac{T_0}{\alpha} \tag{16}$$

where α is:

$$\alpha = \sqrt{1-\beta^2} \tag{17}$$

If v<c, the expression $\sqrt{D_1^2 + v^2 \cdot T_0^2 - 2 \cdot D_1 \cdot v \cdot T_0 \cdot \cos\varphi} \to D$ and the relation (8) can be approximated by the expression (18), (8=>18):

$$\frac{\gamma_0}{\gamma} = \frac{T}{T_0} = 1 - \beta \cdot \cos\varphi + \beta \cdot \frac{v \cdot T_0}{2 \cdot D_1} \tag{18}$$

The distance D (D_1) can take different values for the same frequency γ_0 (One can't determine D from 8 or 18; D is indeterminate. Practically, the frequency γ is a real function of γ_0 and β; γ is a function of γ_0, T_0, or $\lambda_0 = c \cdot T_0$; The distance D can't takes any value; It must be a multiple of λ_0). The relation (18) takes mandatory the forms (19) for a quantum distance ($D_1 = n \cdot c \cdot T_0$) and (20) when n takes mandatory the basic value (n=1) to keep the own original wave (one utilize just the basic frequency for n=1, see the final relation 20; for other frequencies then we can already speak about other waves):

$$\frac{\gamma_0}{\gamma} = \frac{T}{T_0} = 1 - \beta \cdot \cos\varphi + \frac{1}{2} \cdot \beta^2 \cdot \frac{1}{n} \tag{19}$$

$$\frac{\gamma_0}{\gamma} = \frac{T}{T_0} = 1 - \beta \cdot \cos\varphi + \frac{1}{2} \cdot \beta^2 \tag{20}$$

First, the relation (20) can be utilized to determine the period T when one know the source period T_0 and the source velocity, v ($\beta = \frac{v}{c}$); one can speak now about a quantum Doppler Effect relation (20).

Second, if one know the two frequencies (γ, γ_0), one can determine the source velocity v in relation of the observer (β and v=β.c), with the new relation (20) or more rapidly with the classical form (10).

Conclusion

In this chapter one proposes to exchange the classical relation (10) (see [1], p. 114) with the new and more exactly relation (20).

Bibliography

[1] **Bărbulescu N.**, "*Bazele fizice ale relativității Einsteiniene*". Editura Științifică și Enciclopedică, București, 1979, p. 142-148;
[2] **David Halliday, Robert, R.**, - *Physics, Part II,* Edit. John Wiley & Sons, Inc. - New York, London, Sydney, 1966;
[3] **Petrescu-Prahova, M., Petrescu-Prahova, I.**, - *Fizica-Manual pentru anul IV liceu, secția reală,* Editura Didactică și Pedagogică, București, 1976.

CHAPTER III - The Future Energy

Introduction

In the years 70-80 (1970-1980) it foreshadows a serious energy crisis with the rapid depletion of known reserves of oil and gas. The consequences would be catastrophic for mankind, but fortunately came just in time energy produced by nuclear fission. With Nuclear power we have saved, so they were a necessary evil.

Another 2-3 cycles (a cycle is about 30-40 years) they could be useful (even if they will evolve and will use the energy produced by fusion, in which case their effectiveness will increase considerably). However, must to prepare from time, the new energy of the future, the mankind future energy.

The most elegant solution which can be now seen is solar energy. This is practically inexhaustible, in quantities much greater than the planet needs, it is the clean, handy and can become the most affordable (if panels with photovoltaic cells will be produced in industrial quantities increasing).

For make this method to obtaining solar energy, to be totally clean, it is absolutely imperative that converted solar energy into electricity to be distributed directly to national energy networks, to avoid the use of different batteries (polluting).

Although it is small, efficiency of energy conversion (in cells), has increased and will increase further due to scientific research in the field.

It must be made the indication that, all living matter on Earth is the energy of the sun, either directly or indirectly.

The discovery of real particles faster than the speed of light (probably smaller than those known today) may open new chapters in the development of mankind, first in the energy field. Going deeper into matter, and by passing from the quantum level to the sub quantum level, or maybe even deeper, it will determine the increasing energy.

Matter is structured in such a way as if we can penetrating more inside it, the particles of which is composed are increasingly smaller and lighter, more dynamic and more energetic.

Although particle mass decreases, the speed is much greater, so the particle energy is much higher (the energy increases with the mass and with the velocity squared of the particle).

Links to quantum levels (within the atom) are more powerful than the chemical-molecular (in the molecule or between atoms), but lower than those of sub quantum level (in the atomic nucleus, between nucleons), which in turn are overshadowed by the level immediately below, the sub-sub quantum level (in nucleon, between the particles that compose it), and so on until we reach the basic level at which the particle can no longer be divided into other components. If the binding energy is higher, the energy released or required to break or compose these connections, is greater as well.

Hydrogen, as a key component, can be obtained in multiple ways, from almost any item, by nuclear reactions, by the decomposition of water under the action of radiation, by electrolysis of water, etc.

Burning hydrogen it is not a real source of energy (as on Earth, the hydrogen element is not found so much in isolated forms that can be extracted directly and then used as fuel; hydrogen element is generally achieved with energy consumption greater than the energy released by burning it); but it is more a strategic fuel, like a fuel which can be the life-long of internal combustion engines when the oil fuel will lessen or even will disappear.

Wind energy does not represent a real alternative energy, but in some cases it may be a component to complete certain energy goals.

The energy produced from thermal springs in some areas of the planet is very useful, but are very little compared with the needs of the earth.

Probably wave energy of seas and oceans has not given good results since it was not extended and imposed, the more so as we have a planet covered with water at the rate of 70%.

Maybe in the future, the man will exploit the temperature difference between the different levels of seas and oceans, to produce such energy (energies from seas water).

For now, the water remains a serious source of energy in the chapter, hydropower. From water, it extracts the hydrogen, which through burning turns back into water. From water it obtains "heavy water" (by the converting of the element Hydrogen, into heavy isotope named Deuterium, which contains in nucleus in addition to a proton and a neutron), and which is used as nuclear fuel, in some nuclear power plant.

If we look, retrospect and global, the water and the sun are the major energy sources of our planet. Even the living matter (including man), represents a very high proportion water. The water intervenes directly or indirectly in several ways, into the cellular level processes.

Obtaining Energy by the Annihilation of an Electron with a Positron, or Annihilation of a Proton with an Antiproton (case studies presentation)

Getting energy, renewable, clean, friendly (not dangerous), cheaper, by annihilation (For example, the annihilation of an electron with an anti electron). Electron and positron are obtained by extracting them from atoms; the extraction, consume a negligible amount of energy.

Then, the two particles are brought near one another (collision); now it occur the phenomenon of annihilation, when the rest mass is converted totally into energy (gamma photons).

Occur gamma photons, as many as needed to retrieve the total energy of the electron and positron (rest energy and kinetic energy); usually one can get two or three gamma particles (when we have a lower annihilation, ie two antiparticles with lower energy, each with a little beyond rest mass, ie the particles are accelerated at a low-speed motion), but we can get more particles when we have a high annihilation (ie when the particle energy is high and the particles were strongly accelerated before the collision).

Rest energy of an electron-positron pairs exceeds slightly 1 MeV (what is an extremely large energy from some as small particles, comparable energy with that achieved by the merger of two much larger particles, having rest mass of about 2000 times higher).

Hence the first great advantage of the new method proposed, namely that if the most complex physical phenomenon so far tried to get inside the material energy (hot or cold fusion), draw only about a thousandth part of the rest mass of the particle, resulting in the fusion of two particles practically only the energy gap between energy particles being free and their energy when they are united, the proposed method to extract virtually all the internal energy of the particles annihilated.

We started with the electron positron pair because these small particles are more easily extracted from the atoms (the atoms are then immediately regenerated naturally, which determines the nature of renewable energy from the annihilation of particles).

Next step is to test the annihilation between a proton and an antiproton, because their mass is about 1800 times higher than that of the electron and positron, resulting in their annihilation as an energy by about 1000 times higher, ie instead of 1 MeV, 1 GeV (is considered as the only real obtained energy, the energy donated by the proton of the hydrogen ion; but the energy of an antiproton is considered to be donated by us almost entirely, for now, because to obtain today an antiproton we must accelerate some particles at very high-energy and then collide them).

So the real comparison must to be made between the deuterons fusion and annihilation process of a hydrogen ion (proton) with an antiproton. It will be a difference of energy of about 1000 times higher per pair of particles used, in favor of the annihilation process.

Practically it realizes the dream of extracting energy from all the matter. Another great advantage of this method is that no radioactive substances and are not radioactive wastes from the process. From this process we obtain only gamma photons (ie energy) and possibly other energetic mini particles. The process does not pose any threat to humans and the environment.

The energy produced is clean. The technology required is much simpler than nuclear (fission or fusion), cheaper and easier to maintain. Enough energy is given by the annihilation process (virtually unlimited), cheap, clean, safe, renewable immediately (sustainable), with technology made simple.

We can extract the energy of the rest mass of an electron. For a pair of an electron and a positron this energy is circa 1 MeV.

The "synchrotron radiation (synchrotron light source)" produces deliberated a radiation source. Electrons are accelerated to high speeds in several stages to achieve a final energy (that is typically in the GeV range). We need two synchrotrons, a synchrotron for electrons and another who accelerates positrons. The particles must to be collided, after they are being accelerated to an optimal energy level. All the energies are collected at the exit of the Synchrotrons, after the collision of the opposite particles. We will recover the accelerating energy, and in addition we also collect the rest energy of the electrons and positrons.

At a rate of 10^{19} electrons/s we obtain an energy of about 7 GWh / year, if even are produced only half of the possible collisions. This high rate can be obtained with 60 pulses per minute and 10^{19} electrons per pulse, or with 600 pulses per minute and 10^{18} electrons per pulse. If we increase the flow rate of 1,000 times, we can have a power of about 7 TWh / year. This type of energy can be a complement of the fusion energy, and together they must replace the energy obtained by burning hydrocarbons.

Advantages of the annihilation of an electron with a positron, compared with the nuclear fission reactors, are disposal of radioactive waste, of the risk of explosion and of the chain reaction.

Energy from the rest mass of the electron is more easily controlled compared with the fusion reaction, cold or hot.

Now, we don't need of enriched radioactive fuel (as in nuclear fission case), by deuterium, lithium and of accelerated neutrons (like in the cold fusion), of huge temperatures and pressures (as in the hot fusion), etc.

Results and Discussion

How much energy, can we get from inside of the matter? Einstein has showed that from one kg of matter we could get the energy needs for entire Earth for a year:

$$E = m \cdot c^2 = 1[kg] \cdot (3 \cdot 10^8)^2 [(m/s)^2] = 9 \cdot 10^{16} [j] = 2{,}5 \cdot 10^{10} [KWh] = 2{,}5 \cdot 10^7 [MWh] = 2{,}5 \cdot 10^4 [GWh] = 25 [TWh]$$

We could do this, but only if we could extract all the energy from inside the matter.

Through nuclear fusion reaction can be extracted only a part of the rest energy of the particles used. This drop of energy (1 / 1000 of the mass energy of a proton-neutron pairs) is called, discrepancy.

For a kg of particles proton-neutron pairs, fusion energy is about a thousand times smaller than the total energy of a kilogram of matter (only 29 [GWh] from the total internal energy, 25 [TWh]); and considering that a return of 100% fusion reaction, which can't be done anyway.

Theoretically speaking, we can't draw from within the matter (through nuclear fusion reaction) than at most the thousandth part of its energy. Having in view the yield of the nuclear fusion reaction, this obtained energy is and less.

Through reaction of nuclear fission, the energies obtained will be even smaller.

The solution proposed in this work, obtaining energy by the mutual annihilation of two opposite particles, makes possible the requirement of extracting whole energy contained in matter.

A pair formed by a particle and its antiparticle, are brought side by side, at a distance which allow the process of reciprocal annihilation.

To increase the yield of the annihilation reaction (the number of annihilated particles from all particles that exist), we can accelerate the particles and antiparticles separately, and then we may send them into a room where they encounter annihilation at speeds and energies high, or at velocities and energies very high.

If we use electrons and positrons for the reaction of annihilation, it results photons of the gamma type.

In this case, to prevent the possible decay of the obtained photons, again into electrons and positrons (for beginning of this annihilation process with success), the antiparticles and particles used in the process of annihilation, should be collided at low speeds and with low energy.

We can test then the optimum energy particle which permits the reaction with the maxim yield. It is necessary that most particles and antiparticles used, to meet and annihilate each other, and it should be stable as many of the obtained gamma particles.

Conclusions

The fission energy was a necessary evil. In this mode it stretched the oil life, avoiding an energy crisis. Even so, the energy obtained from hydrocarbons represents today about 66% of all energy used. At this rate of use of oil, it will be consumed in about 40 years. Today, the production of energy obtained by nuclear fusion is not yet perfect prepared. But time passes quickly. We must rush to implement of the additional sources of energy already known, but and find new energy sources. In these conditions the proposed method to obtaining energy by annihilation of matter and antimatter, can be a real alternative sources of renewable energy.

References:

[1] EWEA Executive summary "Analysis of Wind Energy in the EU-25" (PDF). European Wind Energy Association.
http://www.ewea.org/fileadmin/ewea_documents/documents/publications/WETF/Facts_Summary.pdf EWEA Executive summary. Retrieved 2007-03-11.

[2] Massachusetts Institute of Technology (2010, September 13). Funneling solar energy: Antenna made of carbon nanotubes could make photovoltaic cells more efficient. *Science Daily*. Retrieved September 21, 2010, from http://www.sciencedaily.com/releases/2010/09/100912151548.htm

[3] "Towards Sustainable Production and Use of Resources: Assessing Biofuels". United Nations Environment Programme. 2009-10-16.
http://www.unep.fr/scp/rpanel/pdf/Assessing_Biofuels_Full_Report.pdf. Retrieved 2009-10-24.

[4] Petrescu, F. New Aircraft. COMEC 2009, Braşov, ROMANIA, 2009.

CHAPTER IV - NEW AIRCRAFT

4.1. ION THRUSTER

4.1.1. About the ion thruster

Speaking about a new ionic engine means to speak about a new aircraft.

The chapter presents shortly the actual ionic engines (called ion thrusters) and the new ionic (pulse) engines proposed by the author.

Ionic engine (ion thruster, which accelerates the positive ions through a potential difference) is about 10 times more effective than classic system based on combustion.

We can still improve the efficiency of 10-50 times if one uses pulses of positive ions accelerated in a cyclotron mounted on the ship; the efficiency can easily grow for 1000 times if the positive ions will be accelerated in a high energy synchrotron, synchrocyclotron or isochronous cyclotron (1-100 GeV). In this, the big classic synchrotron is reduced to a ring surface (magnetic core).

Future (ionic) engine will have mandatory a circular particle accelerator (high or very high energy).

We can thus increase the speed and autonomy of the ship using a less quantity of fuel and power.

One can use synchrotron radiation (synchrotron light, high intensity beams), like high intensity (X-ray or Gamma ray) radiation, as well. In this case will be a beam engine (not an ionic engine), it'll use only the power (energy, which can be solar energy, nuclear energy, or both) and so we will remove the fuel.

It proposes using a powerful LINAC at the exit of synchrotron (especially when one accelerates electrons) to not lose energy by photons premature emission.

With a new ionic engine one builds a new aircraft, which can travel through water and. This new aircraft will can accelerate directly, without an additional combustion engine and without gravity assists from other planets [1].

An *ion thruster* is a form of electric propulsion used for spacecraft propulsion that creates thrust by accelerating ions. Ion thrusters are characterized by how they accelerate the ions, using either electrostatic or electromagnetic force.

Electrostatic ion thrusters use the Coulomb Force and accelerate the ions in the direction of the electric field. Electromagnetic ion thrusters use the Lorentz Force to accelerate the ions. Note that the term "ion thruster" frequently denotes the electrostatic or gridded ion thrusters, only.

The thrust created in ion thrusters is very small compared to conventional chemical rockets, but a very high specific impulse, or propellant efficiency, is obtained.

Due to their relatively high power needs, given the specific power of power supplies, and the requirement of an environment void of other ionized particles, ion thrust propulsion currently is only practicable in outer space.

The first experiments with ion thrusters were carried out by Robert Goddard at Clark College from 1916-1917. The technique was recommended for near-vacuum conditions at high altitude, but thrust was demonstrated with ionized air streams at atmospheric pressure. The idea appeared again in Hermann Oberth's "Wege zur Raumschiffahrt" (Ways to Spaceflight), published in 1923.

A working ion thruster was built by Harold R. Kaufman in 1959 at the NASA Glenn facilities. It was similar to the general design of a gridded electrostatic ion thruster with mercury as its fuel. Suborbital tests of the engine followed during the 1960s and in 1964 the engine was sent into a suborbital flight aboard the Space Electric Rocket Test 1 (SERT 1). It successfully operated for the planned 31 minutes before falling back to Earth.

4.1.2. Hall effect thruster

The Hall effect thruster was studied independently in the U.S. and the USSR in the 1950s and 60s. However, the concept of a Hall thruster was only developed into an efficient propulsion device in the former Soviet Union, whereas in the U.S., scientists focused instead on developing gridded ion thrusters.

Hall effect thrusters were operated on Soviet satellites since 1972. Until the 1990s they were mainly used for satellite stabilization in North-South and in East-West directions. Some 100-200 engines completed their mission on Soviet and Russian satellites until the late 1990s. Soviet thruster design was introduced to the West in 1992 after a team of electric propulsion specialists, under the support of the Ballistic Missile Defense Organization, visited Soviet laboratories.

Ion thrusters utilize beams of ions (electrically charged atoms or molecules) to create thrust in accordance with Newton's third law. The method of accelerating the ions varies, but all designs take advantage of the charge/mass ratio of the ions. This ratio means that relatively small potential differences can create very high exhaust velocities. This reduces the amount of reaction mass or fuel required, but increases the amount of specific power required compared to chemical rockets. Ion thrusters are therefore able to achieve extremely high specific impulses. The drawback of the low thrust is low spacecraft acceleration because the mass of current electric power units is directly correlated with the amount of power given. This low thrust makes ion thrusters unsuited for launching spacecraft into orbit, but they are ideal for in-space propulsion applications.

Hall effect thrusters accelerate ions with the use of an electric potential maintained between a cylindrical anode and a negatively charged plasma which forms the cathode. The bulk of the propellant (typically xenon or bismuth gas) is introduced near the anode, where it becomes ionized, and the ions are attracted towards the cathode, they accelerate towards and through it, picking up electrons as they leave to neutralize the beam and leave the thruster at high velocity.

The anode is at one end of a cylindrical tube, and in the center is a spike which is wound to produce a radial magnetic field between it and the surrounding tube. The ions are largely unaffected by the magnetic field, since they are too massive. However, the electrons produced near the end of the spike to create the cathode are far more affected and are trapped by the magnetic field, and held in place by their attraction to the anode. Some of the electrons spiral down towards the anode, circulating around the spike in a Hall current. When they reach the anode they impact the uncharged propellant and cause it to be ionized, before finally reaching the anode and closing the circuit.

4.1.3. Gridded electrostatic ion thrusters

Gridded electrostatic ion thrusters commonly utilize xenon gas. This gas has no charge and is ionized by bombarding it with energetic electrons. These electrons can be provided from a hot cathode filament and accelerated in the electrical field of the cathode fall to the anode (Kaufman type ion thruster). Alternatively, the electrons can be accelerated by the oscillating electric field induced by an alternating magnetic field of a coil, which results in a self-sustaining discharge and omits any cathode (radiofrequency ion thruster).

The positively charged ions are extracted by an extraction system consisting of 2 or 3 multi-aperture grids. After entering the grid system via the plasma sheath the ions are accelerated due to the potential difference between the first and second grid (named screen and accelerator grid) to the final ion energy of typically 1-2 keV, thereby generating the thrust.

Ion thrusters emit a beam of positive charged xenon ions only. In order to avoid the charging-up of the spacecraft another cathode, placed near the engine, emits additional electrons (basically the electron current is the same as the ion current) into the ion beam. This also prevents the beam of ions from returning to the spacecraft and thereby cancelling the thrust.

Gridded electrostatic ion thruster research (past/present):

- NASA Solar electric propulsion Technology Application Readiness (NSTAR)
- NASA's Evolutionary Xenon Thruster (NEXT)
- Nuclear Electric Xenon Ion System (NEXIS)
- High Power Electric Propulsion (HiPEP)
- EADS Radio-Frequency Ion Thruster (RIT)
- Dual-Stage 4-Grid (DS4G)

4.1.4. Field Emission Electric Propulsion

Field Emission Electric Propulsion (FEEP) thrusters use a very simple system of accelerating liquid metal ions to create thrust. Most designs use either cesium or indium as the propellant. The design consists of a small propellant reservoir that stores the liquid metal, a very small slit that the liquid flows through, and then the accelerator ring.

Cesium and indium are used due to their high atomic weights, low ionization potentials, and low melting points. Once the liquid metal reaches the inside of the slit in the emitter, an electric field applied between the emitter and the accelerator ring causes the liquid metal to become unstable and ionize.

This creates a positive ion, which can then be accelerated in the electric field created by the emitter and the accelerator ring. These positively charged ions are then neutralized by an external source of electrons in order to prevent charging of the spacecraft hull.

4.1.5. Pulsed Inductive Thrusters

Pulsed Inductive Thrusters (PIT) use pulses of thrust instead of one continuous thrust, and have the ability to run on power levels in the order of Megawatts (MW).

PITs consist of a large coil encircling a cone shaped tube that emits the propellant gas. Ammonia is the gas commonly used in PIT engines.

For each pulse of thrust the PIT gives, a large charge first builds up in a group of capacitors behind the coil and is then released. This creates a current that moves circularly. The current then creates a magnetic field in the outward radial direction (Br), which then creates a current in the ammonia gas that has just been released in the opposite direction of the original current.

This opposite current ionizes the ammonia and these positively charged ions are accelerated away from the PIT engine due to the electric field crossing with the magnetic field Br, which is due to the Lorentz Force.

4.1.6. Magnetoplasmadynamic

Magnetoplasmadynamic (MPD) thrusters and Lithium Lorentz Force Accelerator (LiLFA) thrusters use roughly the same idea with the LiLFA thruster building off of the MPD thruster.

Hydrogen, argon, ammonia, and nitrogen gas can be used as propellant. The gas first enters the main chamber where it is ionized into plasma by the electric field between the anode and the cathode. This plasma then conducts electricity between the anode and the cathode.

This new current creates a magnetic field around the cathode which crosses with the electric field, thereby accelerating the plasma due to the Lorentz Force. The LiLFA thruster uses the same general idea as the MPD thruster, except for two main differences.

The first difference is that the LiLFA uses lithium vapor, which has the advantage of being able to be stored as a solid.

The other difference is that the cathode is replaced by multiple smaller cathode rods packed into a hollow cathode tube.

The cathode in the MPD thruster is easily corroded due to constant contact with the plasma. In the LiLFA thruster the lithium vapor is injected into the hollow cathode and is not ionized to its plasma form/corrode the cathode rods until it exits the tube.

The plasma is then accelerated using the same Lorentz Force.

4.1.7. Electrodeless Plasma Thrusters

Electrodeless Plasma Thrusters have two unique features, the removal of the anode and cathode electrodes and the ability to throttle the engine.

The removal of the electrodes takes away the factor of erosion which limits lifetime on other ion engines. Neutral gas is first ionized by electromagnetic waves and then transferred to another chamber where it is accelerated by an oscillating electric and magnetic field, also known as the ponderomotive force.

This separation of the ionization and acceleration stage give at the engine the ability to throttle the speed of propellant flow, which then changes the thrust magnitude and specific impulse values [1].

4.1.8. Plasma Micro Thruster

In the picture number 1 one presents „A Plasma Micro Thruster" Schematic and Prototype (see Figure 1, and [2]).

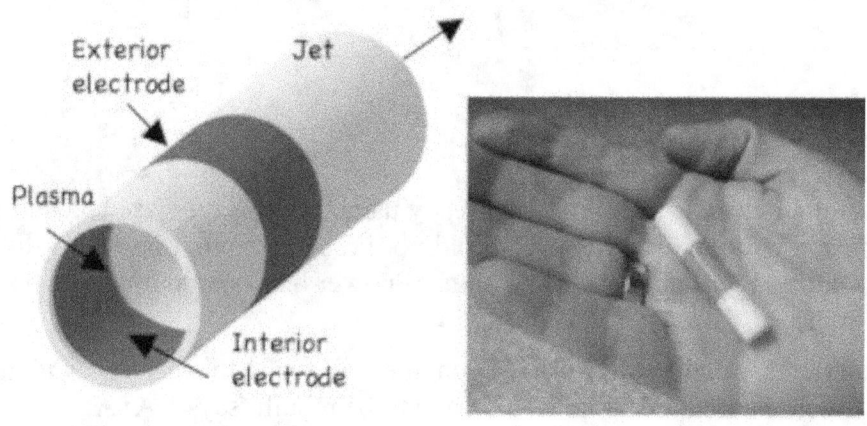

Fig. 1: *Plasma Micro Thruster, Schematic and Prototype*

4.2. THE HIPEP ENGINE

4.2.1. Powerful ion engine relies on microwaves

A powerful new ion propulsion system has been successfully ground-tested by NASA. The High Power Electric Propulsion ion engine trial marks the "first measurable milestone" for the ambitious $3 billion Project Prometheus, says director Alan Newhouse.

The HiPEP engine is the first tested propulsion technology with the potential power and longevity to thrust spacecraft as far as Jupiter without gravity assists from other planets.

These assists involve slingshot maneuvers around planets and can boost the speed of craft significantly. But they require specific planetary alignments, meaning suitable launch dates are rare.

In contrast, a probe powered by a HiPEP engine could launch any time. One goal of Project Prometheus, formerly called the Nuclear Systems Initiative, is to launch a spacecraft towards Jupiter by 2011. The flight would take at least eight years.

The key elements of the HiPEP engine are a high exhaust velocity, a microwave-based method for producing ions that performs for longer than existing technologies and a rectangular design that can more easily be scaled up than circular ones.

Spacecraft are increasingly being built with ion engines rather than engines that burn rocket fuel. This is because ion engines produce more power for a given amount of propellant, and provide a smooth output rather than intermittent spurts.

"Jupiter is such a far away target. Using a chemical system, you just couldn't do it," says John Foster, one of the principal creators of the engine at NASA's Glenn Research Center in Cleveland, Ohio.

The HiPEP engine differs from earlier ion engines, such as that powering NASA's Deep Space One mission, because the xenon ions are produced using a combination of microwaves and spinning magnets. Previously the electrons required were provided by a cathode. Using microwaves significantly reduces the wear and tear on the engine by avoiding any contact between the speeding ions and the electron source.

4.2.2. Nuclear fission

A Japanese asteroid-chasing spacecraft is already using microwave-based technology to produce ions, but Hayabusa uses a small device that could not produce enough power to fly to Jupiter. The HiPEP engine is currently capable of 12 kilowatts of power but its output will be boosted to at least 50 kW for the Jupiter mission.

The rectangular cross section of the HiPEP engine will make this easier, as it can be expanded along one of its sides. A circular engine would have to be rebuilt, says NASA.

Nonetheless, other researchers at NASA's Jet Propulsion Laboratory in Pasadena, California, are working on a cylindrical high-power ion engine, also for the Prometheus project. But Newhouse notes that building a powerful, long-lasting propulsion system is just "one of the pieces we need to get to Jupiter". The electricity for the ion engine is slated to come from on-board nuclear fission reactor. This part of the Prometheus Project is just beginning, with safety considerations, the miniaturization of the reactor and the identity of the fuel all needing to be decided.

4.3. NEW IONIC OR BEAM PULSES ENGINES

By this chapter the author proposes a new pulse engine which works with beam or ionic (ionic beam) pulses.

With a new ionic engine one builds a new aircraft (a new ship). The principal characteristic of this kind of engine is the high power (energy) which accelerates the beam at very high energy, in circular accelerators, in modern linear accelerators (LINAC), or in both.

One can use accelerators similar with the static physics accelerators (synchrotron, synchrocyclotron or isochronous cyclotron).

Ionic engine (ion thruster, which accelerates the positive ions through a potential difference) is about 10 times more effective than classic system based on combustion.

We can still improve the efficiency of 10-50 times if one uses positive ions accelerated in a cyclotron mounted on the ship; the efficiency can easily grow for 1000 times if the positive ions will be accelerated in a high energy synchrotron, synchrocyclotron or isochronous cyclotron (1-100 GeV).

Future (ionic) engine will have mandatory a circular particle accelerator (high or very high energy; see the Figure 3).

Sure that the difficulties will arise from design, but they need to be resolved step by step.

We can thus increase the speed and autonomy of the ship using a less quantity of fuel.

One can use synchrotron radiation (synchrotron light, high intensity beams), like high intensity (X-ray or Gamma ray) radiation, as well. In this case will be a beam engine (not an ionic engine).

A linear particle accelerator (also called a LINAC) is an electrical device for the acceleration of subatomic particles. This sort of particle accelerator has many applications. It used recently as to an injector into a higher energy synchrotron at a dedicated experimental particle physics laboratory. In this, the big classic synchrotron is reduced to a ring surface (magnetic core).

The design of a LINAC depends on the type of particle that is being accelerated: electron, proton or ion.

It proposes using a powerful LINAC at the exit of synchrotron (especially when one accelerates electrons) to not lose energy by photons premature emission (figure 3).

One can use a LINAC in the entry in the Synchrotron and one at out (Figure 2). To use a small entrance LINAC, between him and synchrotron, one put an additional speed circuit in a stadium form (Fig. 2).

The end LINAC can be reduced if one put more end LINACs. See diagram below (Fig. 2.).

Fig. 2: *A high energy synchrotron schema*

This ship has two circular particle accelerators (two synchrotrons)

This ship has first a circular particle accelerator (a synchrotron), and at the end two big linear particle accelerators (two big LINAC)

Fig. 3: *Some flying synchrotron prototypes*

CONCLUSIONS

Speaking about a new ionic engine means to speak about a new aircraft.

The chapter presents shortly the actual ionic engines (called ion thrusters) and the new ionic (pulse) engines proposed by the author. Ionic engine (ion thruster, which accelerates the positive ions through a potential difference) is about 10 times more effective than classic system based on combustion.

We can still improve the efficiency of 10-50 times if one uses pulses of positive ions accelerated in a cyclotron mounted on the ship; the efficiency can easily grow for 1000 times if the positive ions will be accelerated in a high energy synchrotron, synchrocyclotron or isochronous cyclotron (1-100 GeV). Future (ionic) engine will have mandatory a circular particle accelerator (high or very high energy). We can thus increase the speed and autonomy of the ship using a less quantity of fuel and power. One can use synchrotron radiation (synchrotron light, high intensity beams), like high intensity (X-ray or Gamma ray) radiation, as well. In this case will be a beam engine (not an ionic engine), it'll use only the power (energy, which can be solar energy, nuclear energy, or both) and so we will remove the fuel.

A linear particle accelerator (also called a LINAC) is an electrical device for the acceleration of subatomic particles. This sort of particle accelerator has many applications. It used recently as to an injector into a higher energy synchrotron at a dedicated experimental particle physics laboratory. In this, the big classic synchrotron is reduced to a ring surface (magnetic core).

The design of a LINAC depends on the type of particle that is being accelerated: electron, proton or ion.

It proposes using a powerful LINAC at the exit of synchrotron (especially when one accelerates electrons) to not lose energy by photons premature emission (figure 3).

One can use a LINAC in the entry in the Synchrotron and one at out (figure 2). To use a small entrance LINAC, between him and synchrotron, one put an additional speed circuit in a stadium form (fig. 2). With a new ionic engine one builds a new aircraft, which can travel through water and. This new aircraft will can accelerate directly, without an additional combustion engine and without gravity assists from other planets.

Ionic engine (ion thruster) has 2 major advantages (a) and 2 disadvantages (b) compared with chemical propulsion; (a) the impulse and energy per unit of fuel used are much higher; 1-the increased impulse generates a higher speed (velocity; so we can walk longer distances in a short time), 2-the high energy decreases fuel consumption and increase the autonomy of the ship; (b) generate force and acceleration are very small; we can't defeat any forces of resistance to lodging by atmosphere and we have no chance to exceed gravitational forces - ship will not leave a planet (or fall on it) using the ion thruster (It required an additional motor). Vacuum ship acceleration is possible but only with very small acceleration. Increasing more the energy (and also the impulse) can reach the necessary forces and acceleration (Growth will need to be very high, 100 PeV-1000 PeV). Particles energy increased can be made with accelerators circular and or modern linear. Particles energy increased will be huge and in addition will need to grow and the flow of accelerated particles (and the tor diameter; if one increases enough the flow, the necessary energy will be 10 GeV-10 TeV).

Immediate consequence of increasing particle energy will be the increasing of speeds and autonomy of the ship. Now we can achieve huge speeds in a very short time. The ship will pass through any atmosphere (including water) with great ease. The ship can take off or land directly.

Initially one can use to ship the old forms (the old design) which adapts and the accelerator(s).

REFERENCES

[1] Wikipedia, *the free encyclopedia*, net,
[2] Dan Tanna, *Technology today*, edit on 10-6-2008, a net Link.

4.4. Calculation of the momentum of particle jets

$$m = m_0 \cdot \frac{1}{\sqrt{1-\frac{v^2}{c^2}}} = \frac{m_0 \cdot c}{\sqrt{c^2-v^2}} \quad \text{Lorentz}$$

$$\frac{dm}{dv} = \frac{m \cdot v}{c^2 - v^2}$$

$$E_C = \frac{1}{2} \cdot m \cdot v^2$$

$$p = \frac{dE_c}{dv} = \frac{dm}{dv} \cdot \frac{v^2}{2} + \frac{m}{2} \cdot \frac{dv^2}{dv} = \frac{m \cdot v \cdot (2 \cdot c^2 - v^2)}{(2 \cdot c^2 - 2 \cdot v^2)} \Rightarrow$$

$$\Rightarrow p = \frac{m_0 \cdot c \cdot v \cdot (2 \cdot c^2 - v^2)}{2 \cdot \sqrt{c^2-v^2} \cdot (c^2-v^2)} = \frac{m_0 \cdot c \cdot v \cdot (2 \cdot c^2 - v^2)}{2 \cdot (c^2-v^2)^{3/2}}$$

$$\Rightarrow \begin{cases} p = \dfrac{m_0 \cdot c \cdot v \cdot (2 \cdot c^2 - v^2)}{2 \cdot (c^2-v^2)^{3/2}} & \text{when } v \neq c \\ p = \dfrac{h}{\lambda} & \text{when } v \equiv c \end{cases}$$

$\Rightarrow k = M \cdot n \cdot N \cdot p$ Momentum of particle jets

n = Number of pulses per second

N = average number of particles per pulse

M = number of ship engines

$$V_s \cdot M_s = k \Rightarrow V_s = \frac{k}{M_s} \quad \begin{cases} V_s = \text{the speed of the ship} \\ M_s = \text{the mass of the ship} \end{cases}$$

4.5. Calculation of the acceleration of the ship

$$\begin{cases} \dfrac{dp}{dt} = \dfrac{3p \cdot (c^2 - v^2)^{1/2} \cdot v + m_0 \cdot c \cdot (c^2 - \dfrac{3}{2}v^2)}{(c^2 - v^2)^{3/2}} \cdot \dfrac{dv}{dt} & \text{when } v \neq c \\ \dfrac{dp}{dt} = \dfrac{h}{c} \cdot \dfrac{dv}{dt} & \text{when } v \equiv c \end{cases}$$

$$\begin{cases} F = \dfrac{dk}{dt} = M \cdot n \cdot N \cdot \dfrac{dp}{dt} \\ F = M_s \cdot a_s \end{cases} \Rightarrow a_s = \dfrac{M \cdot n \cdot N}{M_s} \cdot \dfrac{dp}{dt}$$

Condition of the ship output from Earth to space

$$a_s \geq g \Rightarrow \dfrac{M \cdot n \cdot N}{M_s} \cdot \dfrac{dp}{dt} \geq g$$

CHAPTER V - CAPTURING ENERGY CONCENTRATED NEAR THE SOURCE AND FORWARDING DIRECTLY TO EARTH IN CONCENTRATED FORM

CAPTURING ENERGY CONCENTRATED NEAR THE SUN

Should start some spatial projects, to capture a large amount of energy somewhere near the source (near the Sun), energy which can be sent then to the Earth in a concentrated form (LASER, MASER, IRASER, etc).

The enormous energy emanating from the sun is spreading in all directions of the universe, and dilute with the distance.

On Earth no longer reach than a small amount from the energy emanated by the sun.

We try here (on the Earth) to capture a drop from a very small amount of energy, who came from Sun. And we also complain that the yield is low, and technological costs are high.

In the next figure we can see how a large amount of energy is transmitted to long distances with low losses, naturally, because is emitted by a sun (a star) in concentrated form, with natural lasers (Figure 1).

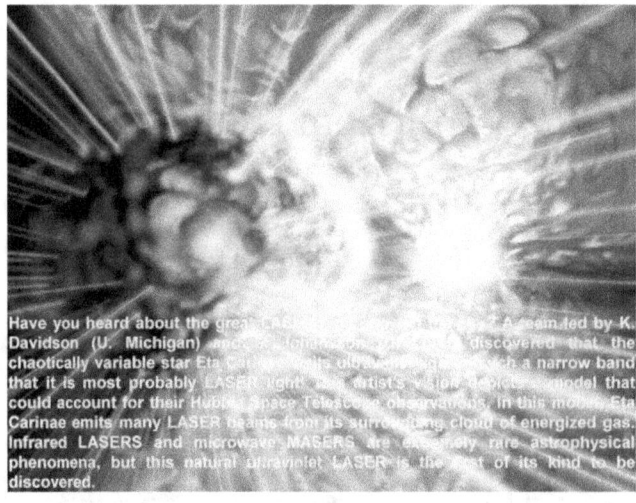

Fig. 1. *A sun that emits radiations LASER*

This is exactly what should we do. This sun strange and extremely rare in Universe, shows us what must we do.

In the next figure (Figure 2) we can see the exact position of our planet in our solar system.

It can see as well how the sun's energy is diluted when the distance from sun grows.

The third halo surrounds the planets Mercury and Venus, and barely touching the Earth (Figure 2).

The fourth halo (the most pale from those which are visible with the naked eye) reach Jupiter.

Mercury is hot, and Saturn is cold.

Fig. 2. *The third halo surrounds the planets Mercury and Venus, and barely touching the Earth*

Installations which must do capturing the solar energy, could be installed over the Mercury.

From the Mercury, the concentrated energy will be transmitted directly focused on the Moon.

On the Moon, the energy will be conserved and forwarded to Earth in doses non-hazardous (with lower concentrations), using multi-channels microwaves.

www.ingramcontent.com/pod-product-compliance
Lightning Source LLC
Chambersburg PA
CBHW081815220526
45470CB00007B/2327